Sundials

Design, Construction, and Use

Denis Savoie

Sundials

Design, Construction, and Use

Illustrations by Thomas Haessig

 Springer

Published in association with
Praxis Publishing
Chichester, UK

Dr Denis Savoie
Département Astronomie-Astrophysique
Palais de la Découverte
Avenue Franklin D. Roosevelt
75008 Paris
France

Original French edition: *Les cadrans solaires*
Published by © Éditions Belin – Paris 2003
Ouvrage publié avec le concours du Ministère français chargé de la culture – Centre national du livre
This work has been published with the help of the French Ministère de la Culture – Centre National du Livre

Translator: Bob Mizon, 38 The Vineries, Colehill, Wimborne, Dorset, UK

SPRINGER–PRAXIS BOOKS IN POPULAR ASTRONOMY
SUBJECT *ADVISORY EDITOR*: John Mason B.Sc., M.Sc., Ph.D.

ISBN: 978-0-387-09801-2 Springer Berlin Heidelberg New York

Springer is a part of Springer Science + Business Media (*springer.com*)

Library of Congress Control Number: 2008934753

Cover design: Jim Wilkie
Translation editor: Dr John W. Mason
Project management: Originator Publishing Services, Great Yarmouth, Norfolk

Printed in Germany

Contents

Illustrations

Preface

*They will merely scorn sundials: no longer used, of course. But they are,
notwithstanding their obvious social role over many centuries, an excellent exercise
for the student who takes no pleasure in verbiage. So, for heaven's sake, start by
learning the theory of sundials! Then you can go on to discuss the history of
astronomical thought to your heart's content! Begin at the beginning: by far the
most difficult course.*

H. Bouasse, *Astronomie Théorique et Pratique*, Paris, 1928

This book is an introduction to the construction of sundials, and to the calculations
involved in that construction. It is aimed at all devotees of astronomy who wish to
know more about sundials. It is also written for teachers of physics and technology,
and for students in teacher training. Those teaching in primary schools will also
find material for simple and amusing applications to do with light and shadow.
Many new educational programmes for schoolchildren may, in fact, contain projects on sundials or on the measurement of time.

Although there are many books on the subject of sundials, they are unfortunately of mixed quality, and often too elementary or too complicated. Some will be
difficult to find, and some are by authors who have not mastered the subject.

Sundials are of course an excellent introduction to astronomy, but first of all it
is necessary to delve into cosmography (the study of the visible universe that
includes geography and astronomy) in order to understand the way they work.
This is the burden of the first chapter, and it may seem hard going compared with
the rest of the book, which is more practical; but useful concepts will have been
established. Too often, this first stage is neglected in works on sundials, meaning
that students cannot absorb all the notions involved.

Moreover, we must recognize that some of those notions are difficult. For
example, it is not immediately obvious why an equatorial sundial has two faces.

And how do we explain to the novice why an analemmatic sundial is so called, and why it is elliptical and its style movable? How can we convey to students the notion of the two components of the equation of time?

Knowing how the Sun moves through the sky in the course of a year as seen from your observing site is the best introduction to the workings of a sundial. The best way to learn about this is undoubtedly by attending planetarium shows: beneath the artificial sky-vault, practically all the movements of the Earth can be simulated. This will increase your understanding of the nature of, for example, the ecliptic, the local meridian, the phenomenon of the seasons, and the inclination of the Earth's axis of rotation. It is so much harder to convey these things on a blackboard!

A sundial is not just a theoretical instrument. Making a working model is a source of great satisfaction, as theory is turned into reality. Often, having set up a sundial and seen it tell the hour and the date, we come to understand certain things which were hitherto unclear.

This book deals only with classic sundials, easy to make from wood or card with compasses, a protractor, a ruler and a calculator. Only plane trigonometry is used, and each formula is demonstrated. Fact boxes, often of a historical nature, occur throughout the chapters, sometimes with an indication of the appropriate academic level, and there are web-based exercises.

With our insistence on the practical and observational aspects of sundials, we hope that readers will be able to create a sundial, and in their own way, using their own particular modifications and improvements. Most of these ideas have been tested out at summer camps, by young people with no previous knowledge of astronomy.

Finally, this book deals with sundials in tropical and southern hemisphere regions, but this is a subject which, if fully explored, would need a book of its own! There is one formula which is not demonstrated here: the equation for the tip of the shadow of a gnomon, since the study of conics belongs to higher and special mathematics. However, in the form in which it is given, it may easily be employed by sixth-form students and may be applied in that form to equatorial, horizontal, polar or vertical sundials in either hemisphere.

For those who wish to go further into the calculations involved in sundials, the bibliography will direct them towards further reading.

I must express my very grateful thanks to two of my colleagues in the Sundial Commission: to Serge Grégori, the most accomplished sundial 'hunter' I know, for supplying the marvellous photos which embellish this book; and to Olivier Escuder, the 'systematist' of sundial mottoes, for his selection of original examples. Finally, I would also like to thank my friend and colleague Marc Goutaudier, of the *Palais de la Découverte*, for his invaluable help. He took on the heavy task of proofreading and checking all the calculations in the original French language edition, and I am very grateful to him.

Denis Savoie
Paris, France
December 2008

1 A little astronomy

A full understanding of the apparent motion of the Sun is the necessary starting point for the understanding of the workings of a sundial. The cosmographical ideas in this chapter will be of use throughout the book, in understanding the characteristics of different types of sundials.

1.1 The moving Earth

It is useful to recall first of all that the Earth moves in three principal ways: it rotates on its axis, it revolves about the Sun, and its rotational axis precesses. We are concerned only with the first two of these motions when discussing sundials.

An observer studying the Earth from a distance of a few million kilometers would see our planet rotating from west to east in about 24 hours. Also, the Earth would be moving around the Sun in the ecliptic plane in a little over 365 days (Figure 1.1). The Earth's axis always leans in the same direction relative to the ecliptic plane. For our external observer, the consequences of these two motions would be obvious: it would soon become apparent that, on Earth, there is a succession of days and nights and a sequence of seasons.

Now, to an observer on Earth, the combination of these two motions is much less obvious. This observer has the impression that the Earth is motionless and that

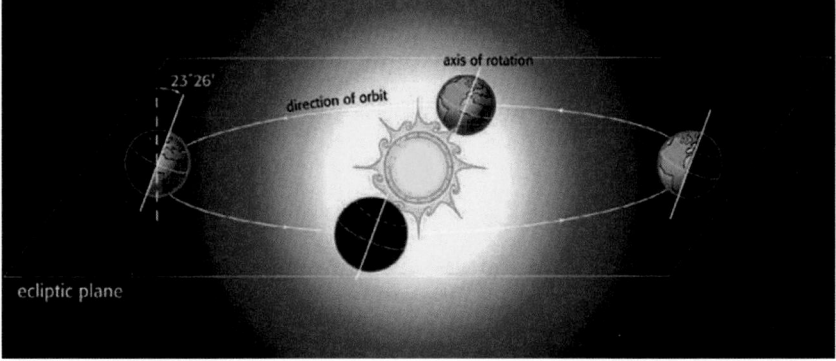

Figure 1.1 The Earth rotates on its axis every 24 hours approximately, and orbits the Sun, in the plane of the ecliptic, in just over 365 days. The angle between the axis of the Earth's rotation and the vertical to the plane of the ecliptic is currently 23° 26′.

the heavenly bodies are moving: so, on Earth, we talk about the apparent motion of the heavens. In particular, if we discuss sundials, we tend to think of the apparent motion of the Sun rather than of the real motion of the Earth.

1.2 Geographical coordinates

The way in which we construct sundials depends on our geographical location. So we must define the various terms used to specify our position on the surface of the Earth.

Parallels and meridians
Assuming the Earth to be perfectly spherical, we first of all determine an imaginary straight line PP′ around which our planet rotates (Figure 1.2). This line is known as the Earth's axis, or its axis of rotation. It encounters the Earth's surface at two points, the North Pole (P) and the South Pole (P′). The longest circular parallel which is perpendicular to the axis of rotation is known as the Equator, dividing the Earth into the northern and southern hemispheres. The smaller circles parallel to the Equator are known as parallels of latitude, the most important among them situated 23° 26′ north of the Equator (the Tropic of Cancer), and 23° 26′ south of the Equator (the Tropic of Capricorn). The importance of these two Tropics derives from the fact that only in locations between these two lesser parallels can the Sun be at the zenith.

The parallels lying at 66° 34′ N and 66° 34′ S are respectively the Arctic and

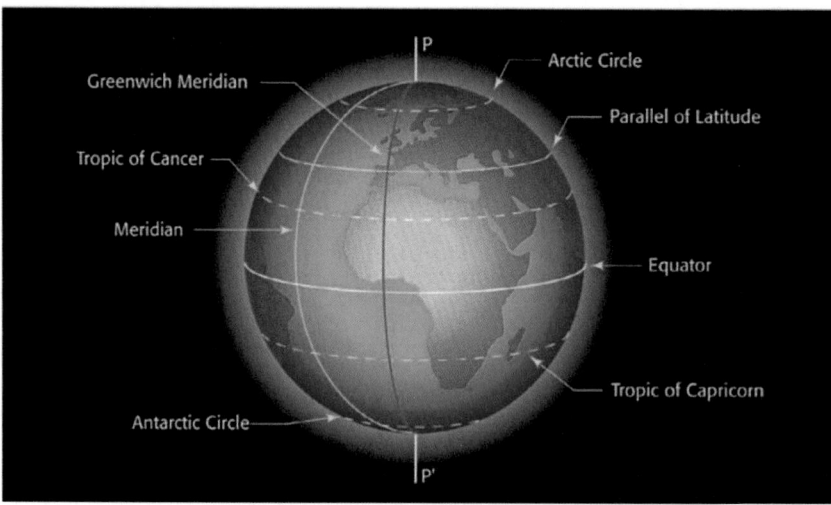

Figure 1.2 The surface of the Earth with its parallels and meridians.

Antarctic Circles. In latitudes greater than these, the "midnight sun" is observed, i.e. the phenomenon of the Sun being above the horizon at local midnight.

The semicircles between the poles are known as meridians. The meridian passing through the Royal Greenwich Observatory is known as the prime (or international) meridian.

Latitude and longitude

In order to describe a location on the Earth's surface, we use geographical coordinates of latitude and longitude (Figure 1.3). The latitude of a place (φ) is the angle between the Equator and the vertical at that place. It is expressed in degrees from the Equator (with positive sign in the northern hemisphere, $0°$ to $+90°$, and negative sign in the southern hemisphere, $0°$ to $-90°$). For example, France extends from around latitude $41° 55'$ N (Ajaccio) to $50° 57'$ N (Calais). One degree of latitude represents approximately 111.11 km on the ground.

The longitude λ of a place is the dihedral angle between the local meridian and the Greenwich Meridian (longitude zero), from $0°$ to $+180°$ westwards, and from $0°$ to $-180°$ eastwards (following the sign convention of astronomy). Longitude is expressed either in degrees, minutes and seconds $(°, ', '')$ or in hours, minutes and seconds (h, m, s). Note that $15° = 1$ h. The longitude of the Paris Observatory is $-2° 20' 15''$ ($2° 20' 15''$ E), or -0 h 9 m 21 s.

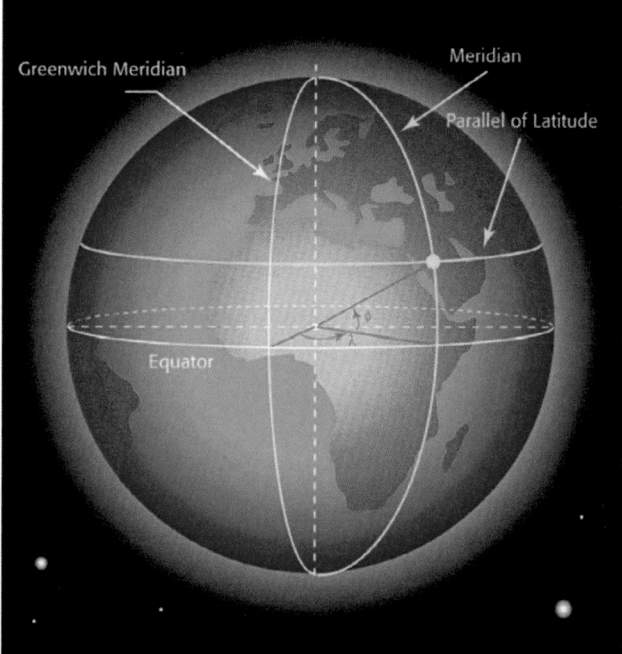

Figure 1.3 Latitude φ and longitude λ of a place on the surface of the Earth. The Equator is the parallel of reference for latitude, and the Greenwich meridian the meridian of reference for longitude.

Further information
● Conversion of degrees into hours: see Appendix G, page 165.

1.3 The Celestial Sphere and the local Celestial Sphere

The Celestial Sphere

For an observer on the Earth's surface, our planet seems motionless while the spherical vault of the heavens seems to move around it. Since the radius of the Earth is negligible compared with the distance to the heavenly bodies, including the Sun, we can consider the Earth to lie exactly at the center of this sphere. Now this imaginary sphere, of great and arbitrary size compared with the size of the Earth, is known as the Celestial Sphere, and it is centered upon the center of the Earth. Observation of the night sky will show that the stars seem to travel from east to west around an imaginary axis which encounters the inner surface of the Celestial Sphere close to the Pole Star (Figure 1.4). We observe exactly the same principle in a planetarium. The imaginary axis is the rotational axis of the Earth, passing through the Celestial Sphere at two opposite points, the North Celestial Pole and the South Celestial Pole. By analogy with the Earth, we also define a Celestial Equator.

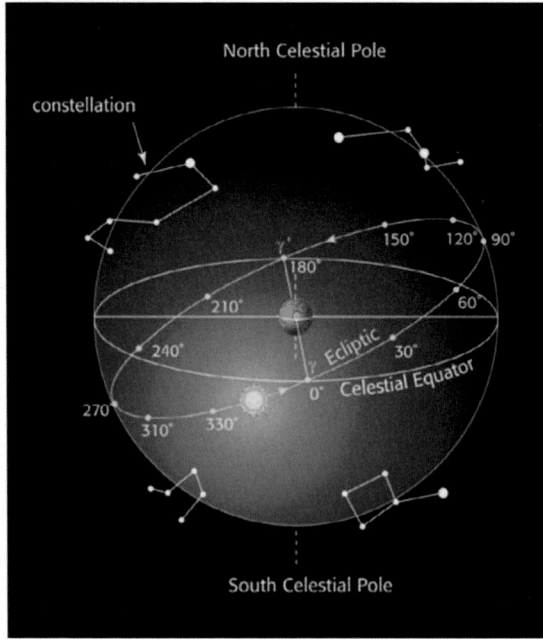

Figure 1.4 The Celestial Sphere.

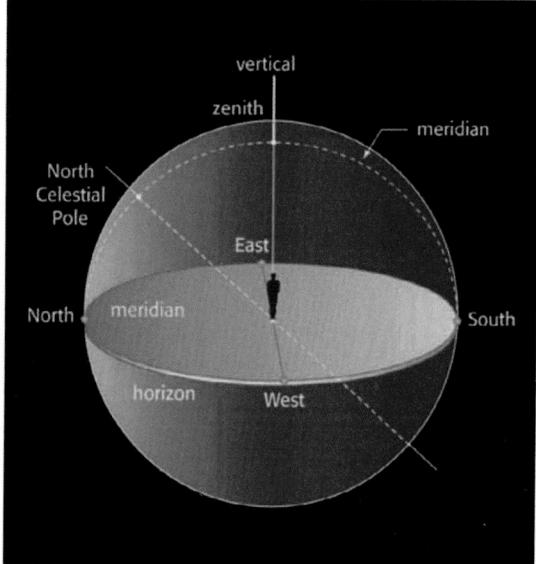

Figure 1.5   The local Celestial Sphere. The Celestial Sphere is referred to when discussing apparent annual motions, while the local Celestial Sphere is used for apparent diurnal and nocturnal paths.

The local observer's Celestial Sphere

For any point on the Earth's surface, we can define a local Celestial Sphere (Figure 1.5). On this sphere, it is possible to represent the movement of heavenly bodies using two fixed references: the vertical at the observer's location, which can be determined by using a plumb line; and the horizon, which is a great circle at right angles to the vertical. It is on this horizon that the cardinal points (north, east, south and west) are defined. The observer's vertical passes through a unique point, the zenith, 90° from the horizon.

Through the North and South Poles passes a great circle, the local meridian. This passes through the zenith of the observer's location and indicates geographical north and south. The altitude in degrees of the North Celestial Pole is the same as the observer's latitude.

1.4   The Sun's annual motion

The ecliptic

Since the publication in 1543 of Nicolaus Copernicus' *De Revolutionibus*, in which he affirmed, among other things, that the Earth rotates and moves around the Sun, it has been recognized that the motion of the 'day-star' is an apparent motion. Not only is the Sun carried from east to west through the sky because of the Earth's rotation (diurnal motion), but it also seems to move throughout the year relative to the stars (annual motion). So, if it were possible to observe the Sun and the stars at the same time, it would become obvious that the Sun moves about one degree per day eastwards against the starry Celestial Sphere. By noting the Sun's position every day, we can verify that, a year later on the same date, it returns to the same point against the backdrop of the stars.

Therefore, in the course of a year, the center of the Sun describes a great circle, known as the ecliptic, on the Celestial Sphere. The Sun's path is so called because solar and lunar eclipses can occur only when the center of the Moon is very near the plane of the ecliptic.

The inclination of the ecliptic relative to the Celestial Equator is known as the **obliquity of the ecliptic** (ε): this is also the angle between the equatorial polar axis PP$'$ and the ecliptic polar axis $\Pi\,\Pi'$ (Figure 1.6). At present, the value of the

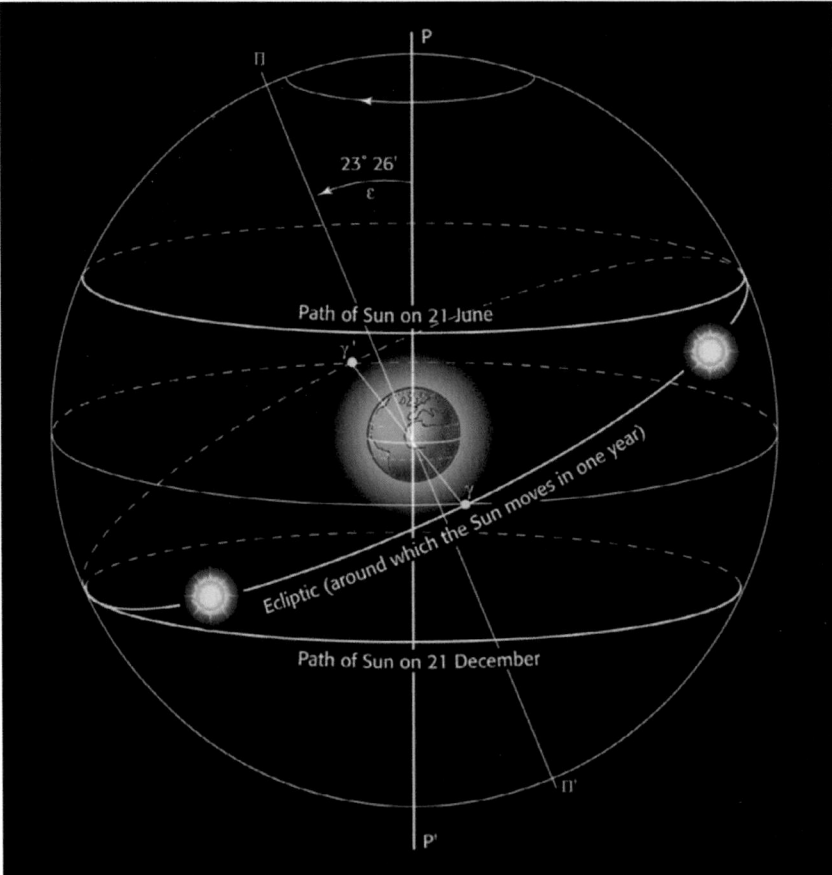

Figure 1.6   The apparent annual path of the Sun across the Celestial Sphere. In the course of a year, the Sun describes a great circle (the ecliptic) on the Celestial Sphere.

obliquity of the ecliptic is $23° \, 26'$, and it is decreasing by about $1'$ per century. The value of $23° \, 27'$ given in many books refers to the obliquity at the beginning of the twentieth century.

Noteworthy points

The ecliptic cuts the Celestial Equator at two points, γ and γ'. These are known as the nodes. γ is a particularly important point, being the ascending node (vernal equinox). The Sun crosses this point at the spring equinox around 20 March, moving from the southern into the northern hemisphere. γ' is the descending node (autumnal equinox). Here, the Sun crosses the ecliptic at the autumnal equinox around 23 September, moving from the northern into the southern hemisphere. The other two notable points on the ecliptic are points E and E', at opposite ends of a diameter perpendicular to the line of the nodes: these are the points of the Sun's passage through the solstices.

The mean interval between two passages of the Sun through the vernal equinox is known as the tropical year, its current value being 365 d 5 h 48 m 45 s.

Note that the dates of the spring equinox, summer solstice, autumnal equinox and winter solstice quoted here, and in subsequent paragraphs, refer to the northern hemisphere. In the southern hemisphere, the seasons are reversed.

Solar longitude

Solar longitude, written as ℓ_0 (not to be confused with the longitude of a place), is the angle between the vernal point γ and the Sun (Figure 1.7). It is measured along the ecliptic in the direct (west-east) sense, and is expressed in degrees from $0°$ to $360°$. The Sun moves approximately one degree eastwards every day along the ecliptic. When the longitude of the Sun is zero, the Sun is at point γ, i.e. the vernal equinox, on 20 March. At $90°$, the Sun reaches its greatest angular distance above the equator ($+23° \, 26'$). This is the summer solstice on 21 June. At $180°$, the Sun crosses point γ'. This is the autumnal equinox, on 23 September. Finally, at $270°$, the Sun reaches its greatest angular distance below the equator ($-23° \, 26'$). This is the winter solstice, on 21 December.

The seasons

In astronomy, a season is the time taken by the Sun to traverse each of the quadrants between points γ, E, γ' and E' (Figure 1.7):

> spring corresponds to the arc γE;
> summer corresponds to the arc $E\gamma'$;
> autumn corresponds to the arc $\gamma'E'$
> winter corresponds to the arc $E'\gamma$.

The seasons are not all of equal length. Currently, the shortest (in the northern hemisphere) is winter and the longest is summer. This inequality is due to the fact that the movement of the Sun in longitude is not uniform. The Sun stays in the northern hemisphere for 186 days, and in the southern for 179 days; in other

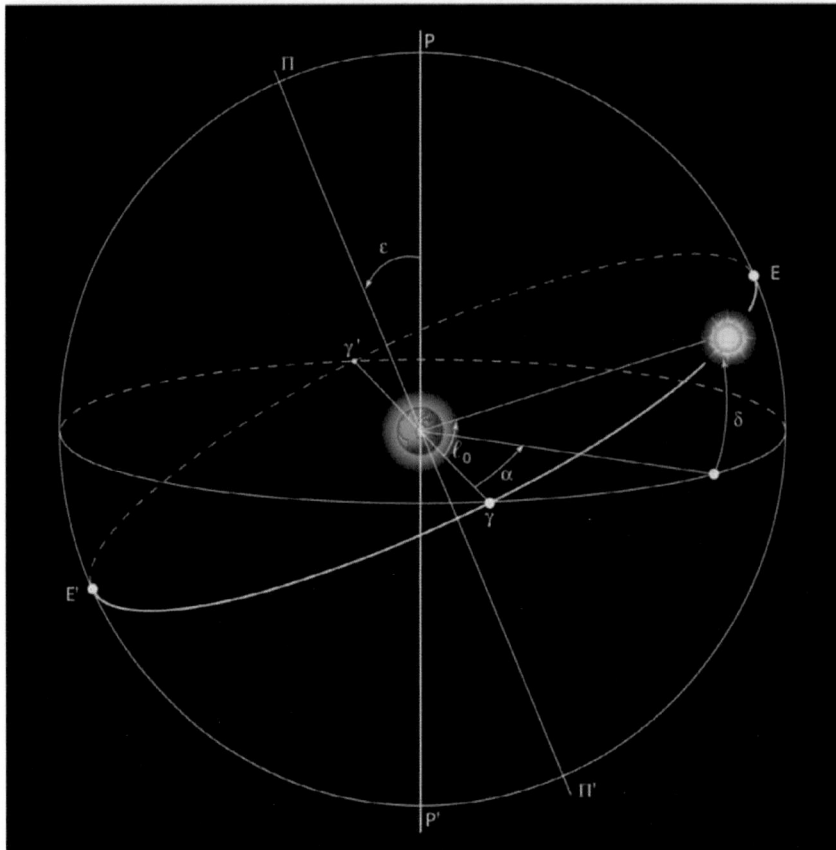

Figure 1.7 Characteristic quantities for the position of the Sun on the ecliptic: longitude ℓ_0, right ascension α and declination δ.

words, in Europe and North America, spring and summer combined are longer than autumn and winter. The reverse is true for locations in the southern hemisphere.

The declination of the Sun

The movement of the Sun in longitude corresponds to a variation in its angular distance from the Equator. This is known as the declination of the Sun, and is written as δ (Figure 1.7). At the spring equinox (20 March) this declination is zero. Its value increases thereafter until the summer solstice of 21 June, to a maximum of $+23° 26'$. Then it decreases to zero at the autumnal equinox (23 September), and the value becomes negative after this date. It reaches a minimum value of $-23° 26'$ on

21 December, at the winter solstice. Lastly, the declination increases, to become zero once again on 20 March.

To sum up, the value for the declination of the Sun increases between the winter solstice and the summer solstice, and decreases between the summer solstice and the winter solstice.

Right ascension

As well as the Sun's movement in declination, there is its movement in right ascension, written as α. This is the projection onto the Celestial Equator of the Sun's longitude. Right ascension is measured in hours from 0 h to 24 h, from point γ, in the direct sense. Right ascension is zero at the spring equinox, 6 h at the summer solstice, 12 h at the autumnal equinox and 18 h at the winter solstice. Table 1.1 brings together the values expressing the position of the Sun at the equinoxes and solstices.

The declination δ of the Sun and its right ascension α are linked to the solar longitude ℓ_0 thus:

$$\tan \alpha = \cos \varepsilon \tan \ell_0$$

$$\sin \delta = \sin \varepsilon \sin \ell_0,$$

ε being the obliquity of the ecliptic. At the end of this book there is a table giving the declination of the Sun for each day of the year at 12 h UT (Appendix C, page 148).

In gnomonics (the study of sundials), it is normally assumed that the Sun's declination does not vary in the course of a day, though in reality there is a slight change in its value. At the equinoxes this variation reaches a maximum value of almost 1' per hour.

Moreover, as the Earth moves round the Sun in one year, with its axis inclined and pointing towards the Pole Star, the altitude of the Sun varies throughout the year as seen from every point on the surface of the globe.

The ecliptic is represented on the Celestial Sphere and not on the local Celestial Sphere (Figure 1.6). In effect, the annual motion of the Sun is independent of its diurnal motion, a fact well worth remembering!

Table 1.1 Values of longitude, right ascension and declination of the Sun at the equinoxes and solstices.

	Spring equinox	Summer solstice	Autumnal equinox	Winter solstice	Spring equinox
Longitude	0°	90°	180°	270°	360°
Right ascension	0 h	6 h	12 h	18 h	24 h
Declination	0°	+23° 26′	0°	−23° 26′	0°

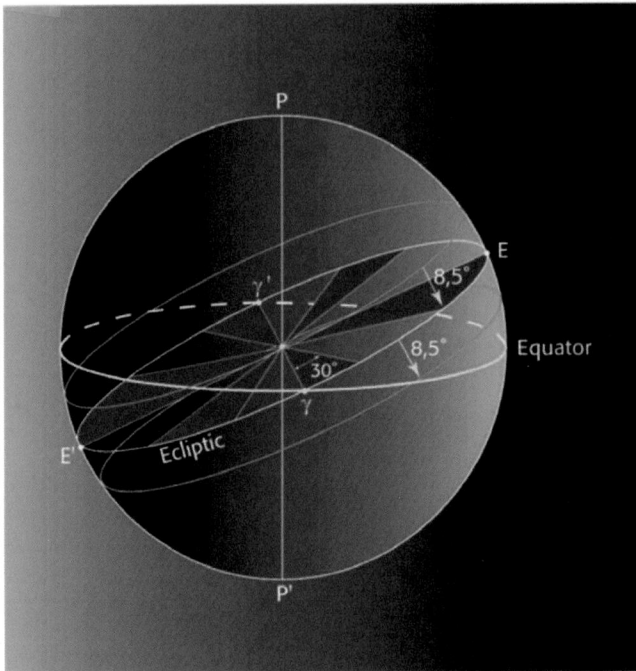

Figure 1.8   The signs of the Zodiac on the Celestial Sphere. Each sign of the Zodiac corresponds to a 30-degree zone of longitude on the ecliptic, and lies between two circles situated on either side of the ecliptic.

The Zodiac

The Zodiac corresponds to a band delimited by two circles parallel to the ecliptic, at an angular distance of $8°\ 30'$ on either side of it. The twelve signs of the Zodiac are Aries (The Ram), Taurus (The Bull), Gemini (The Twins), Cancer (The Crab), Leo (The Lion), Virgo (The Virgin), Libra (The Scales), Scorpio (The Scorpion), Sagittarius (The Archer), Capricorn (The Sea-goat), Aquarius (The Water Bearer) and Pisces (The Fishes). Each Zodiacal sign corresponds to a zone of $30°$ in longitude along the ecliptic (Figure 1.8). The signs of the Zodiac are not to be confused with the Zodiacal constellations themselves. There are twelve signs, but thirteen constellations, twelve of which bear the names of the signs, the other being Ophiuchus. The thirteen constellations divide the ecliptic into unequal parts in longitude.

1.5   The Sun's diurnal motion

Diurnal motion, i.e. the apparent east–west motion of the Sun as the day passes, is the result of the Earth's axial rotation. This motion proceeds from sunrise, through the Sun's meridian passage (true or solar noon), to sunset.

Azimuth

The Sun rises and sets at different points along the horizon according to the seasons. The azimuth A of the Sun is the angle in the horizontal plane between the direction of the Sun at a given moment, and due south (Figure 1.9). Thus, azimuth is reckoned along the horizon from $0°$ to $+180°$ westwards, and from $0°$ to $-180°$ eastwards. The azimuth of the cardinal point due East is $-90°$, and the azimuth of the cardinal point due West is $+90°$.

We can show (Appendix D, page 153) that the azimuth of the Sun at sunrise and sunset is linked to its declination δ and to local latitude φ, by the equation:

$$\cos A = -\sin \delta / \cos \varphi$$

For example, in the case of Paris ($\varphi = 48° \ 50'$), at the time of the summer solstice ($\delta = +23° \ 26'$), we find that $A = -127° \ 10'$ at sunrise and $+127° \ 10'$ at sunset, meaning that the Sun is rising in the north-east and setting in the north-west (Figure 1.10). At the winter solstice ($\delta = -23° \ 26'$), we find that $A = -52° \ 50'$ at sunrise and $+52° \ 50'$ at sunset: the Sun is rising in the south-east and setting in the south-west.

At the equinoxes $A = -90°$ (due east) at sunrise and $+90°$ (due west) at sunset.

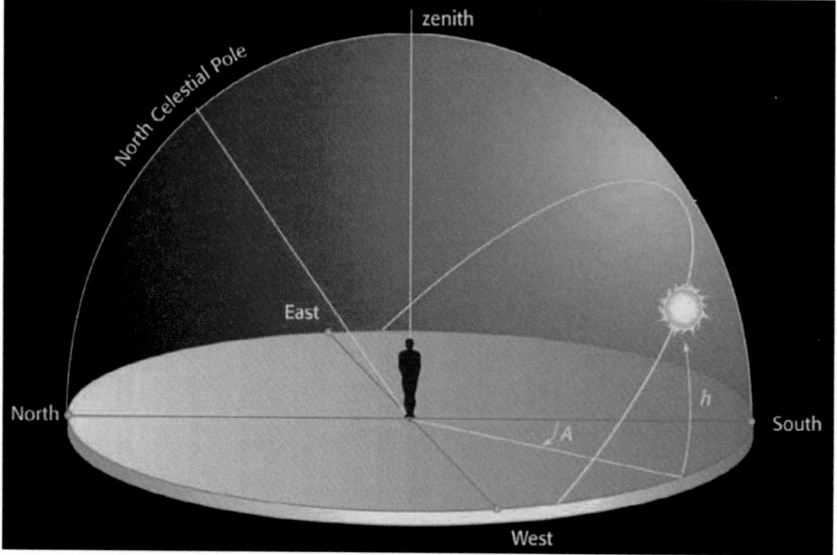

Figure 1.9 The azimuth of the Sun. The azimuth A is the angle between the direction of the Sun, along the horizontal plane, and the southern point. Its value depends on the time of day, the season and the location of the observer.

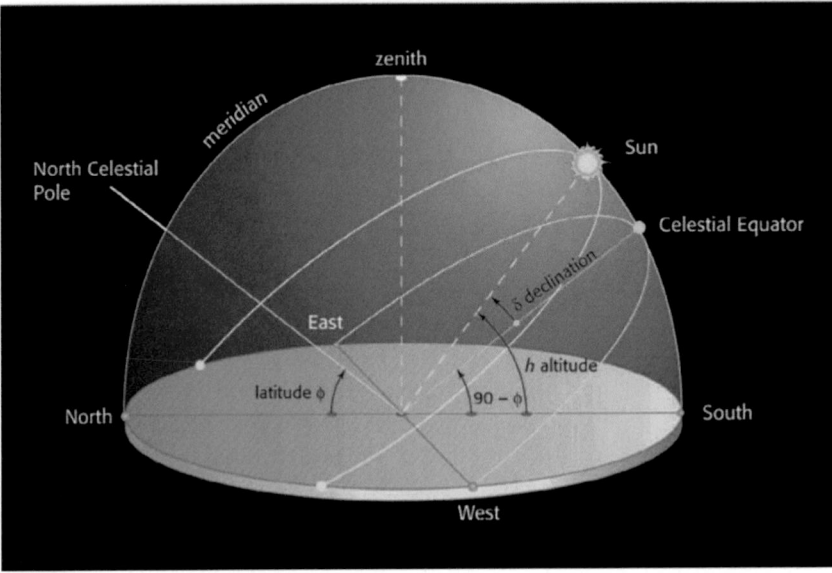

Figure 1.10 The apparent path of the Sun and altitude h. When the Sun is on the local meridian, it culminates due south (in Europe). Its altitude depends on its declination δ and the latitude φ.

In summary, the Sun rises south of east and sets south of west when its declination has a negative value (23 September–20 March). Only at the equinoxes does it rise due east and set due west (though, strictly speaking, this is true only if we ignore the variation on the Sun's declination throughout the day). The Sun rises north of east and sets north of west when its declination has a positive value (20 March–23 September). Note that these remarks are equally valid for the southern hemisphere.

The altitude of the Sun at true noon

When the Sun crosses the meridian, its azimuth is zero. This is known as the culmination of the Sun, and occurs due south (in Europe and North America). The Sun's altitude h at this time can be expressed as a function of local latitude φ and the Sun's declination δ, on that day, thus:

$$h = 90° - φ + δ$$

For example, in Paris ($φ = 48°\ 50'$, $δ = +23°\ 26'$) on June 21, $h = 64°\ 36'$; at the winter solstice, $h = 17°\ 44'$. At the equinoxes, $h = 90° - φ$: for Paris, $41°\ 10'$ (Figure 1.10).

The two relationships shown above become more complicated if we consider some particular moment of the day (see formulae for diurnal motion, Appendix D, page 150).

The Sun at the zenith
At the summer solstice, the Sun passes through the zenith (altitude 90°) for places on the Tropic of Cancer (latitude 23° 26′ N). At the equinoxes, the Sun passes through the zenith for places on the Equator (latitude 0°). At the winter solstice, the Sun passes through the zenith for places on the Tropic of Capricorn (latitude 23° 26′ S). The Sun never passes through the zenith as seen from Europe and North America. In the intertropical zones, the Sun passes through the zenith twice annually at true noon; its declination has the same value as local latitude: $\delta = \varphi$. A rod planted vertically in the ground at these places would have no shadow at solar noon on those two days. In the intertropical regions, the Sun can culminate in both the north and the south.

Day length
Another consequence of the Sun's changing declination is the variation in the length of the day. At European and North American latitudes, the day lasts longer in summer than in winter, a fact easily explained. In summer, the Sun rises in the north-east, climbs high into the sky towards its culmination, then descends towards the north-western horizon. Thus, it describes an arc of considerable length in the sky. In winter, by contrast, the Sun rises in the south-east, culminates at a lower point in the sky, and sets in the south-west. Its diurnal arc is much shorter.

Day length varies not only seasonally, but also as a function of latitude. For example, in spring and summer in the northern hemisphere, the nearer one is to the Pole, the longer the day becomes. On and above the Arctic Circle, a daylight period 24 hours long is attained. There, the Sun, permanently above the horizon, is said to be circumpolar.

At a given location, day length, expressed in degrees, is equal to $2H_0$, where H_0 is the semi-diurnal arc of the Sun (Figure 1.11). It can be shown (see Appendix D, page 150) that H_0 is related to the observer's latitude and to the declination of the Sun on a given date thus:

$$\cos H_0 = -\tan \varphi \tan \delta$$

To obtain the day length in hours, we need only to divide the calculated value of $2H_0$ by 15. For example, for Paris ($\varphi = 48° 50′$) at the summer solstice ($\delta = +23° 26′$), $H_0 = 119° 42′ 54''$. Therefore $2H_0 = 239° 25′ 47''$: in hours 15 h 57 m 43 s, or practically 16 hours. At the winter solstice ($\delta = -23° 26′$), day length is approximately 8 hours, varying therefore by a factor of two in a year.

At the equinoxes, day length is always 12 hours (ignoring the effect of refraction), whatever the location.

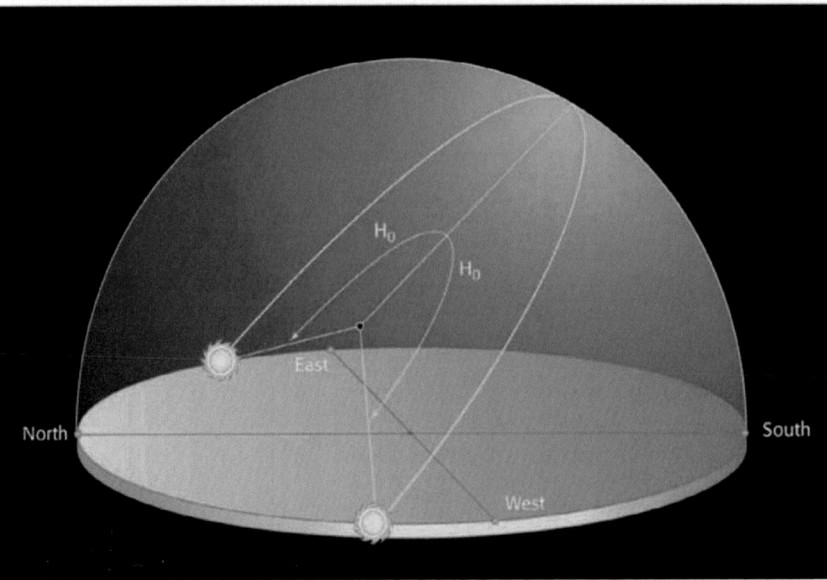

Figure 1.11 The semi-diurnal arc H_0. Day length depends on the location and the date.

At the Equator, day and night are of equal 12-hour length throughout the year (ignoring the effect of refraction). At any other location on Earth, this is not the case, except at the equinoxes. In spring and summer in the northern hemisphere, the day is longer than the night; while in the southern hemisphere, it is shorter (the seasons being reversed in the northern and southern hemispheres).

Calculating sunrise and sunset times

By calculating the semi-diurnal arc H_0, it is possible to calculate the rising and setting times of the Sun, thus:

$$\text{sunrise} = 12\,\text{h} - H_0$$
$$\text{sunset} = 12\,\text{h} + H_0$$

These times are expressed as solar times. To convert them into clock time, we subtract the equation of time E, and add the local longitude λ, and any additional hours appropriate for 'Winter Time' or 'Summer Time'. The aim is not to match the accuracy of times given in the ephemerides. Moreover, we consider here that the declination of the Sun remains invariable during the day, and also, we take no account of atmospheric refraction (which makes the Sun appear higher near the horizon) nor of the Sun's diameter.

1.6   The hour angle of the Sun

Measuring time

When we say that a sundial measures solar time, we are not really telling the truth. In reality, a sundial is measuring an angle: the rotational angle of the Sun in its daily path or trajectory (Figure 1.12). It is this angle that we interpret as a time. Since antiquity, it has been known that this angle does not change at a constant rate during the day as the year progresses. The Sun does not move at a constant 15° every hour: sometimes it travels a little faster, sometimes a little slower. So solar

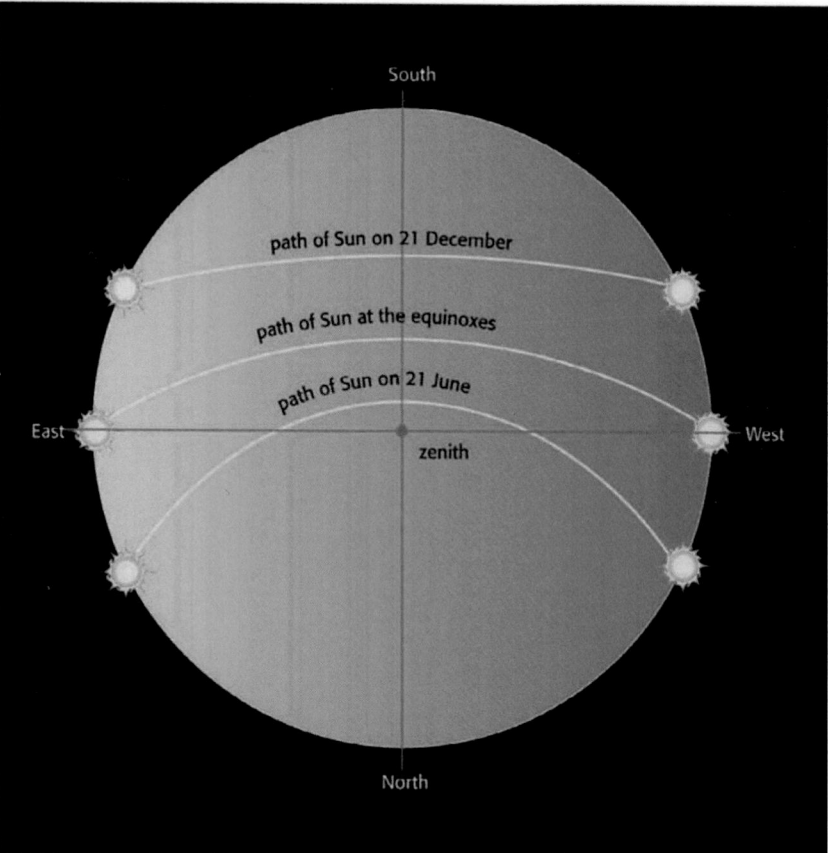

Figure 1.12   The path of the Sun in one day across the Celestial Sphere as seen from the zenith. On December 21, the Sun rises in the south-east and sets in the south-west. At the equinoxes, it rises in the due east and sets due west. On June 21, the Sun rises in the north-east and sets in the north-west.

time as measured by a sundial is not uniform, and we must seek to correct its irregularities.

The greatest of the ancient astronomers, Ptolemy, author in the second century AD of *The Almagest*, one of the most influential astronomical texts of his era, strove to do just that. When Ptolemy made an astronomical observation, he noted the exact moment of the phenomenon in solar time. His goal was to establish a theory of the movement of the wandering celestial bodies (Sun, Moon and planets), in order to be able to predict their future positions: he therefore had to use a uniform time scale in his models. For this reason, Ptolemy corrected for the irregularities in solar time in all his observations, by means of the equation of time. Understanding the Sun's hour angle and why it does not vary in a constant way is undoubtedly the most difficult aspect of this book: however, nineteen centuries ago, these notions were perfectly understood by Ptolemy!

The hour angle of the Sun

In positional astronomy, and in gnomonics, the art of constructing sundials, we use a fundamental quantity: the hour angle of the Sun, known as H (Figure 1.13). This is the angle, measured along the Celestial Equator from the point of culmination of the Sun, between the local meridian and the meridian which passes through the Sun. It is reckoned from $0°$ to $+180°$ westwards, and from $0°$ to $-180°$ eastwards. When the Sun is on the meridian, the value of the hour angle is zero. At sunrise or sunset, the angle is known as semi-diurnal arc H_0, such that $\cos H_0 = -\tan \varphi \tan \delta$.

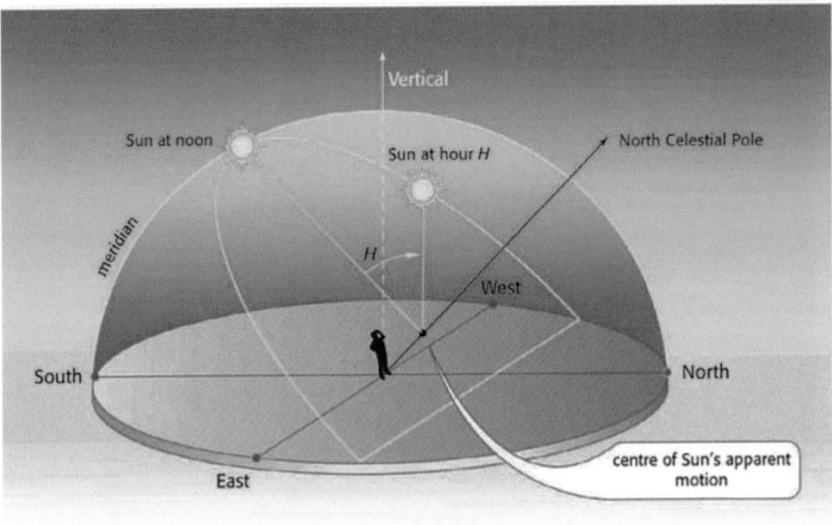

Figure 1.13 The Sun's hour angle H. It is measured along the Celestial Equator from its culmination in the south.

For sundials, we postulate that one hour corresponds to a variation of $15°$ in the hour angle H of the Sun, since the Sun completes its $360°$ journey in 24 hours. It must be remembered that a sundial measures the hour angle of the Sun, and that this angle is expressed as a time (known as true solar time). For example, if we read off a time of 18 h on a sundial, this means that the Sun's hour angle is equal to $+90°$; for 10 h, the angle would be $-30°$.

In reality, the Sun's hour angle involves two motions: diurnal motion, i.e. the motion of the Earth as it turns on its axis; and annual motion, i.e. the apparent eastward displacement of the Sun along the ecliptic.

Further information

- What is the relationship between the hour angle and the azimuth?: see Appendix F, page 160.

1.7 The equation of time

The mean Sun and the true Sun

The time interval between two successive passages of the Sun across the local meridian is known as the true solar day. The solar day is not of constant length, varying in a year between 23 h 59 m 39 s and 24 h 0 m 30 s. It is easy to verify this, as in the example given of meridian passages as seen from Paris on different dates (Table 1.2). In August, for example, the interval measured between the meridian passages on the 13th and 14th of the month is 23 h 59 m 49 s. So, compared with an imaginary Sun which returns to the meridian after exactly 24 h, the Sun is 11 seconds 'fast'. On the day after this, the Sun will be even 'faster', and so on. Since ancient times, we have had the concept of a 'mean Sun', crossing the meridian at

Table 1.2 Times when the Sun crosses the meridian at Paris and lengths of the true solar day for dierent dates.

Date	Instant of Sun's meridian passage at Paris	Interval
13 August 2003	11 h 55 m 34 s	
		23 h 59 m 49 s
14 August 2003	11 h 55 m 23 s	
		23 h 59 m 48 s
15 August 2003	11 h 55 m 11 s	
25 December 2003	11 h 50 m 32 s	
		24 h 0 m 30 s
26 December 2003	11 h 51 m 02 s	
		24 h 0 m 30 s
27 December 2003	11 h 51 m 32 s	

intervals of 24 h 0 m 0 s. This mean Sun therefore moves at a uniform rate along the Celestial Equator, covering 360° in 24 h.

The 'true Sun' corresponds to the real Sun as observed in the sky. The sum of the variations (when 'fast' or 'slow') of the true Sun, as opposed to the mean Sun, is known as the equation of time. The equation of time is a very important consideration for those interested in sundials. It has two astronomical causes, not always easy to comprehend, which we shall now describe.

Reduction to the Equator

We have seen that a sundial measures the hour angle of the Sun, and that the hour angle is reckoned along the Celestial Equator. Now, we have also noted that the Sun is on the Celestial Equator on only two occasions during the year, at the equinoxes; at other times it is either above the equator (positive declination) or below the equator (negative declination).

At a given moment (Figure 1.14), the hour angle of the true Sun is H and the hour angle of the mean Sun is H_m. By definition, the equation of time is equal to $H - H_m$.

In other words, the equation of time represents the difference between *true* time and *mean* solar time.

In order for us be able to compare the two hour angles, they must both be in the same plane, i.e. that of the Equator.

Now, the true Sun is located on the ecliptic, which as we have seen, is inclined to the Equator. We therefore have to project the true Sun onto the Equator. This projection entails our first inequality in solar time, and is known as **reduction to the Equator**.

This inequality has a six-month period and reduces to zero four times a year, at the equinoxes and the solstices. Its present extreme values are −9.87 and +9.87 minutes. This means that the mean Sun and the true Sun may be as much as 9.87 minutes 'adrift', if we ignore the second inequality.

The equation of the center

In the early seventeenth century the German astronomer Kepler established the three laws which bear his name (only the first two concern us here). These deal with the motion of the Earth around the Sun. Kepler's first law states that the Earth describes an ellipse around the Sun of very small eccentricity (0.017), and that the Sun is at one of the foci of this ellipse. For the purposes of demonstration, and to continue in the convention already adopted, we shall assume that it is the Sun which is moving around the Earth. It follows that the distance between the Sun and the Earth changes throughout the year. The Sun is nearest to the Earth in winter, around 3 January (perigee), and furthest away in summer, around 4 July (apogee). The heliocentric (i.e. Sun-centred) model has the Earth at perihelion in winter and at aphelion in summer.

Kepler's second law states that the radius vector between the Sun and the Earth sweeps out equal areas in equal times. For example (Figure 1.15), arc AB is swept out in the same time as arc CD. Since the distance between the Earth and the

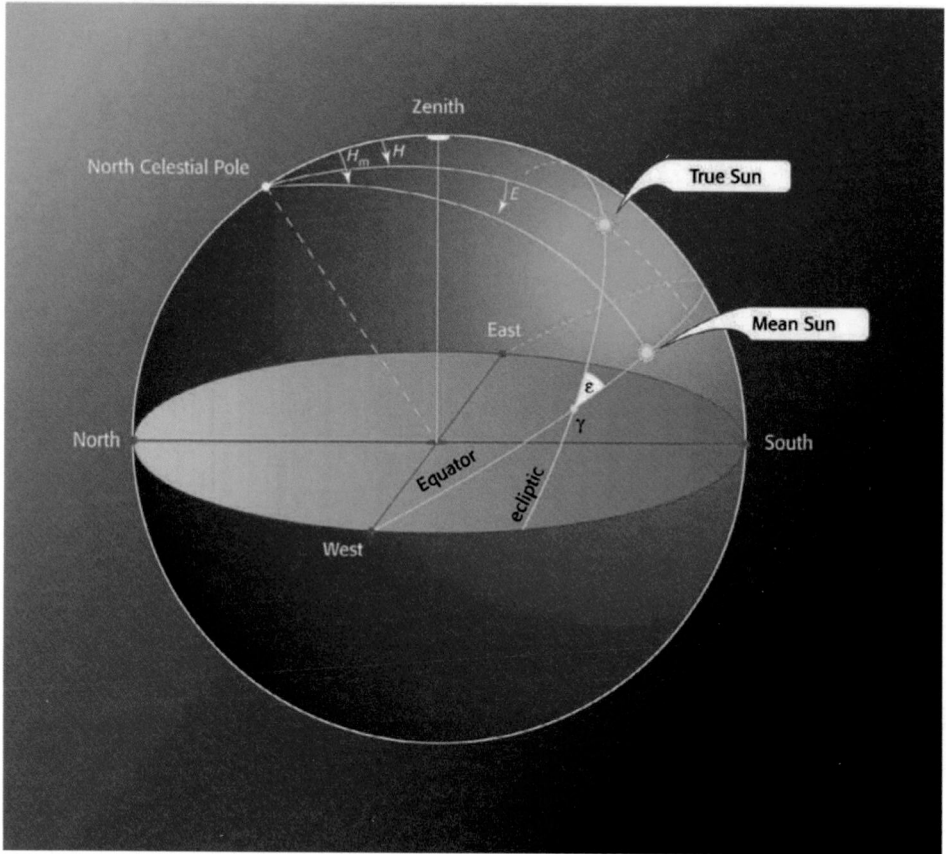

Figure 1.14   Reduction to the Equator. The hour angles of the true Sun and the mean Sun.

Sun varies, the second law implies that the Sun does not travel at the same rate throughout the year, i.e. the Sun's ecliptic longitude does not vary at a uniform rate: the Sun moves faster in winter and more slowly in summer. For example, in January the Sun moves along the ecliptic by $1°\ 01'$ in 24 h, while in July this decreases to $0°\ 57'$ in 24 h. It is for this reason that the seasons are not all of the same length. Remember always that here we are dealing with the revolution of the Earth around the Sun, and not its diurnal motion.

Let us now imagine a fictitious Sun which moves along a circular ecliptic at a uniform rate, i.e. a Sun which travels the $360°$ of the ecliptic in one tropical year (365.2422 days). Needless to say, the real Sun moves sometimes faster, and some-times more slowly, than this fictitious Sun (Figure 1.16). It might be said that the real Sun oscillates to either side of the fictitious Sun. If we call the longitude of the

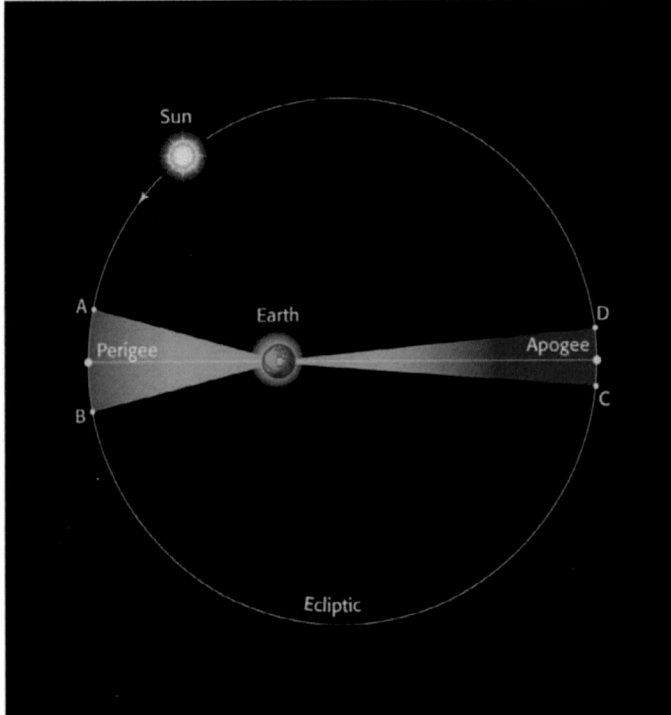

Sun

A
Earth
D
Perigee
Apogee
C
B

Ecliptic

Figure 1.15
Kepler's Second
Law. The arc
AB is swept out
in the same
time as the arc
CD.

fictitious Sun L and that of the real Sun ℓ_0, the difference between these longitudes, expressed in hours, minutes and seconds, is known as the equation of the center, with a value of $\ell_0 - L$. This equation reaches zero twice a year when the Sun is at perigee and apogee. Its current extreme values are −7.66 and +7.66 minutes.

The equation of time
To sum up, true solar time is modified by two inequalities, the first arising from the inclination of the ecliptic to the equator, and the second due to Kepler's laws. The sum of these two inequalities gives us the equation of time. Since these two terms have different amplitudes, different periods, and are out of phase, the resultant curve is not symmetrical with respect to the line representing value zero.

The equation of time may be positive or negative. If it is positive, the true Sun crosses the observer's meridian before the mean Sun. If it is negative, it means that the real Sun crosses the meridian after the mean Sun.

The equation of time has a value of zero four times a year, on 15 April, 13 June, 1 September and 25 December. It is more than 14 minutes 'fast' at the beginning of February and a little less than 16 minutes 'slow' at the beginning of November

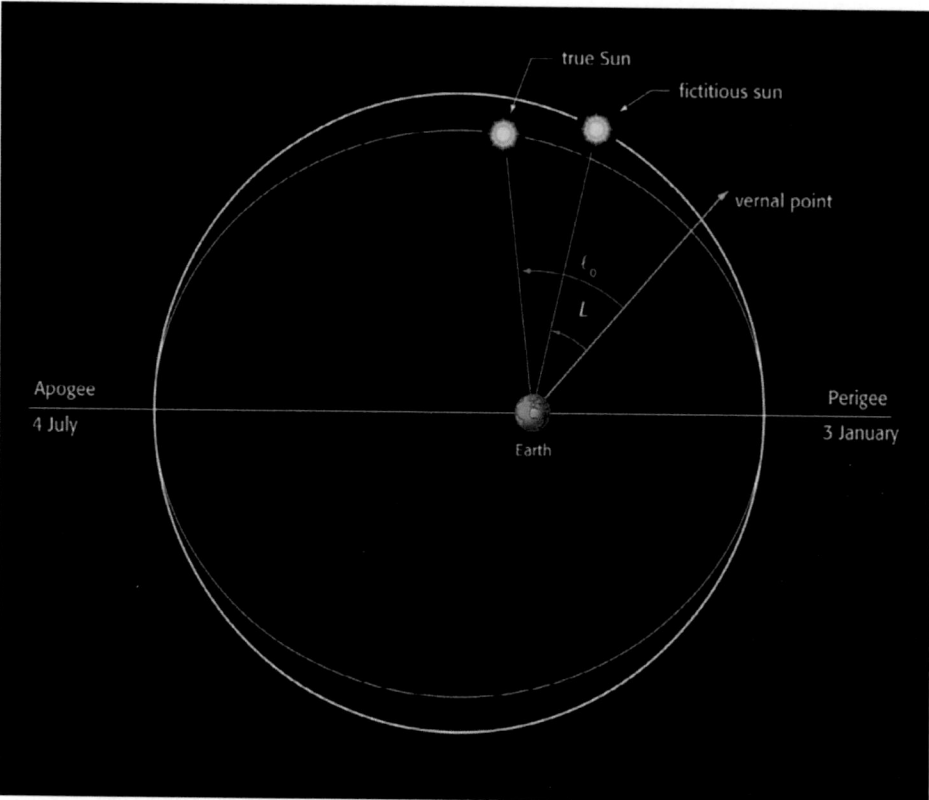

Figure 1.16   Differences in longitudes between the fictitious Sun, with its apparent circular path, and the true Sun, with its apparent elliptical path.

(Figure 1.17). As the years pass, the equation of time slowly evolves due to the variation of the Earth's orbital elements.

Remember also that the variations in the equation of time are accompanied by variations in true day length: around the winter solstice the true day is at its longest (24 h 0 m 30 s); in mid-September, it is at its shortest (23 h 59 m 39 s). It is worth noting that if the Sun did circle the Earth at a uniform rate, the equation of the center would have zero value, whilst the equation of time would not, because of obliquity. To make the equation of time disappear completely, the axis of the Earth would have to become perpendicular to the plane of the ecliptic.

Applications of the equation of time

In France, since the 19th century, clocks have been regulated according to mean time; as we have seen, the difference between the true time of our sundials and the

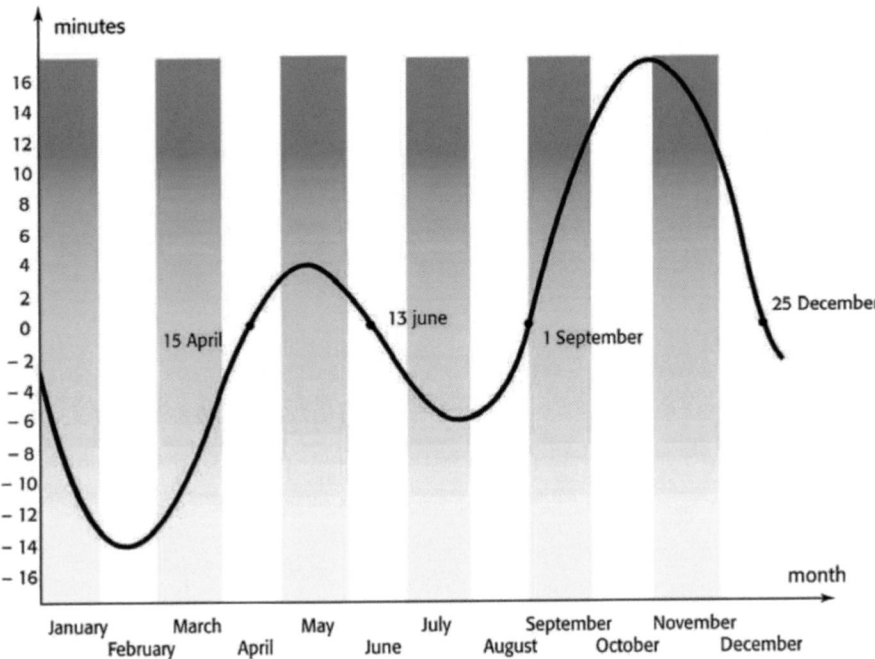

Figure 1.17 The variation of the equation of time in one year.

mean time of our clocks, an abstraction arising from corrections due to longitude and legalities, constitutes what we call the equation of time.

In Figure 1.18, we can trace the evolution of day length for the latitude of Paris around the winter solstice. On 6 December, for example, true noon occurs 9 minutes before mean noon. At Christmas, on 25 December, the two noons coincide. On 7 January, true noon occurs 6 minutes after mean noon. During this period the curves representing sunrise and sunset, which are more or less symmetrical around noon, approach each other, reaching a minimum separation of 8 h 02 m on the day of the winter solstice. Thereafter, they diverge. If our clocks were still set, as they once were, by true solar time, we could verify the fact that the earliest sunsets occur on 21 and 22 December.

The same is not the case with mean time! The earliest sunset corresponds with the date on which the daily increase in the equation of time exactly compensates for the decrease in the half-length of the day.

In Figure 1.18, the distance between mean noon and sunset reaches its minimum value when the tangent to the sunset curve and the noon line become parallel. This is the case on 12 December, and in Paris, this is the day of earliest sunset.

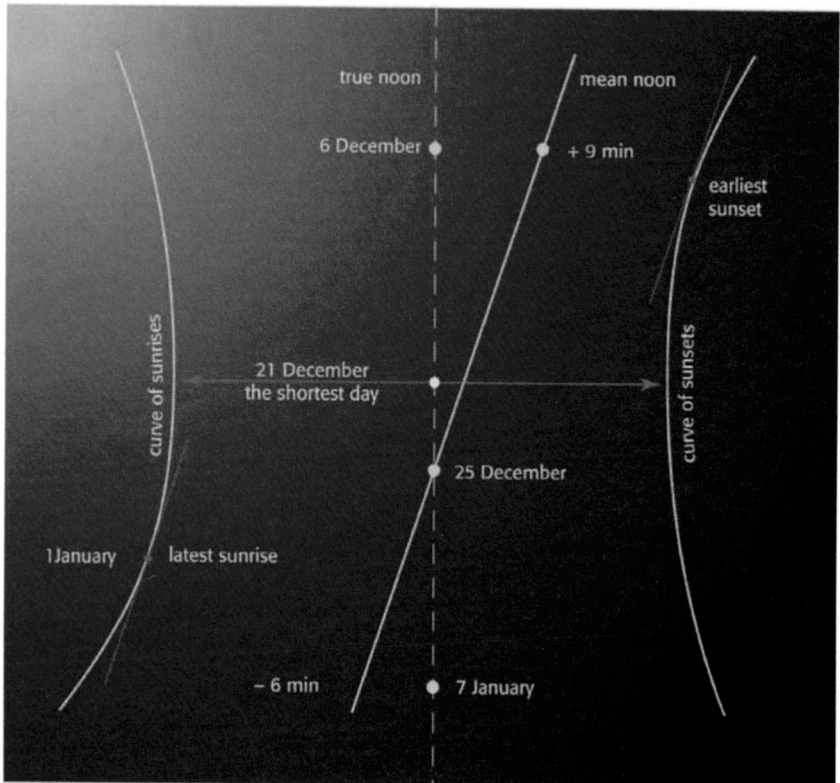

Figure 1.18   The evolution of day length for the latitude of Paris. The difference between mean noon and true noon has been exaggerated.

If we look further into this, we see that the mean noon line continues to approach the sunrise curve. Sunrises, according to our clocks, occur later each day, until 1 January. The equation of time, which does not permit the co-existence of latest sunrises and earliest sunsets at the winter solstice, plays a similar role at the summer solstice too.

Further information
- Meridians and mean solar time: see Appendix E, page 155.

St. Lucy's Day

"A la Sainte Luce, le jour croît du saut d'une puce"

There is an old tradition in France that on 13 December (St. Lucy's Day), the day at last lengthens, by "the hop of a flea", referring to the fact that, from this date, days begin to grow longer. However, if we consult tables of sunrise and sunset times on a calendar or in the ephemerides, we see that this is most certainly untrue! The shortest day of the year is that of the winter solstice around December 21. This means that the latest sunrise and the earliest sunset occur on that day. But … even this is not exactly true, if we re-examine those sunrise and sunset times. Behind all this confusion lurk the reform of the calendar and the equation of time.

Our present calendar is largely based on that established in 45 BC by Julius Caesar and his astronomical advisor, Sosigenes. It was Sosigenes who adopted the value of 365.25 days (365 d 6 h) for the length of the year, and introduced the concept of a leap year falling every four years. The so-called Julian year, named after Julius Caesar, was in fact 11 minutes too long compared with the true length of a year, meaning that the dates of equinoxes and solstices began to 'drift' as the centuries passed. The spring equinox passed from 20 March into 19 March, then 18 March, etc. Similarly, the winter solstice drifted through 19 December, 18 December, etc., and by the fourteenth century, the solstice fell on St. Lucy's Day, 13 December: hence the old saying. In 1582, the reform of the calendar, initiated on the authority of Pope Gregory VIII, moved the date of the winter solstice back to its proper place. St. Lucy no longer has the shortest day, and a little historical research has told us why.

2 An introduction to sundials

Long ago, the observation that the shadow of an object changes during the course of both the day and the year led humankind to the realization that the Sun might be used as a timekeeper.

2.1 A brief history of sundials

A sundial comprises a table, upon which are inscribed hour lines, and a style, which casts a shadow on the table. The shadow moving across the hour lines indicates the time. The first sundial may well have been a gnomon, a stick set vertically in the ground, but we have no idea of when this was first used, or to which civilization we should ascribe its invention. It was probably seen as an astronomical instrument rather than as a timepiece. As the millennia have passed, all sorts of sundials, of many degrees of sophistication, have been introduced. The tables of these sundials may be horizontal, vertical, inclined, and even spherical or cylindrical. They are therefore known as horizontal sundials, vertical sundials, etc. The style may also point in various directions: it may be vertical, horizontal, inclined ..., or may even be replaced by a hole through which a beam of light passes.

The first sundials
Since the time of the ancient Greeks, techniques of sundial construction have been evolving, principally in the various forms of sundials constructed: for example, there have been conical sundials, spherical sundials, etc. All ancient sundials had one thing in common: they indicated hours of variable duration and it was only the tip of the shadow which indicated the time. Hundreds of ancient sundials, for the most part Greek in origin, have been recovered from archaeological excavations throughout the Mediterranean area. Many of these show a high degree of sophistication, suggesting great understanding of mathematics and astronomy; a good example is the Tower of the Winds at the Agora in Athens, with its eight sundials (Figure 2.1). This was designed by Andronicus of Cyrrhus in about 50 BC. We do not know how these dials were marked out, but they remain among the finest examples of Greek gnomonics. Other examples of ancient sundials are shown in Figures 2.2 and 2.3.

Figure 2.1   The Tower of the Winds in Athens (1st century BC). On each side of this octagonal tower there is a sundial. The hour lines are still discernible, but they are not visible on this photograph. (Photo: CORBIS/S.Vannini)

At the time of the Roman conquest, sundials were imported into Italy. In Book VII of his *Natural History*, Pliny the Elder tells a tale well known to experts on ancient Rome: after the capture of Catania in Sicily at the time of the First Punic War (264–241 BC) against Carthage, the first sundial to be brought to Rome was set up on a column in a public square. Pliny states that the divisions on the sundial did not agree with the hours of the day, but the dial was used for 99 years before it was replaced by a more accurate instrument. It seems unlikely that the Romans would have had such blind faith in an inaccurate sundial over such a long period: they probably took little notice of it and continued to study the apparent motion of the Sun. One famous sundial was that installed on the Campus Martius (Field of Mars) in Rome by the Emperor Augustus, to commemorate his victory in Egypt. In 10 BC, he had an obelisk nearly 22 meters high brought from Heliopolis to Rome. Atop the obelisk was a ball designed to counteract penumbral effects. The dial was nearly

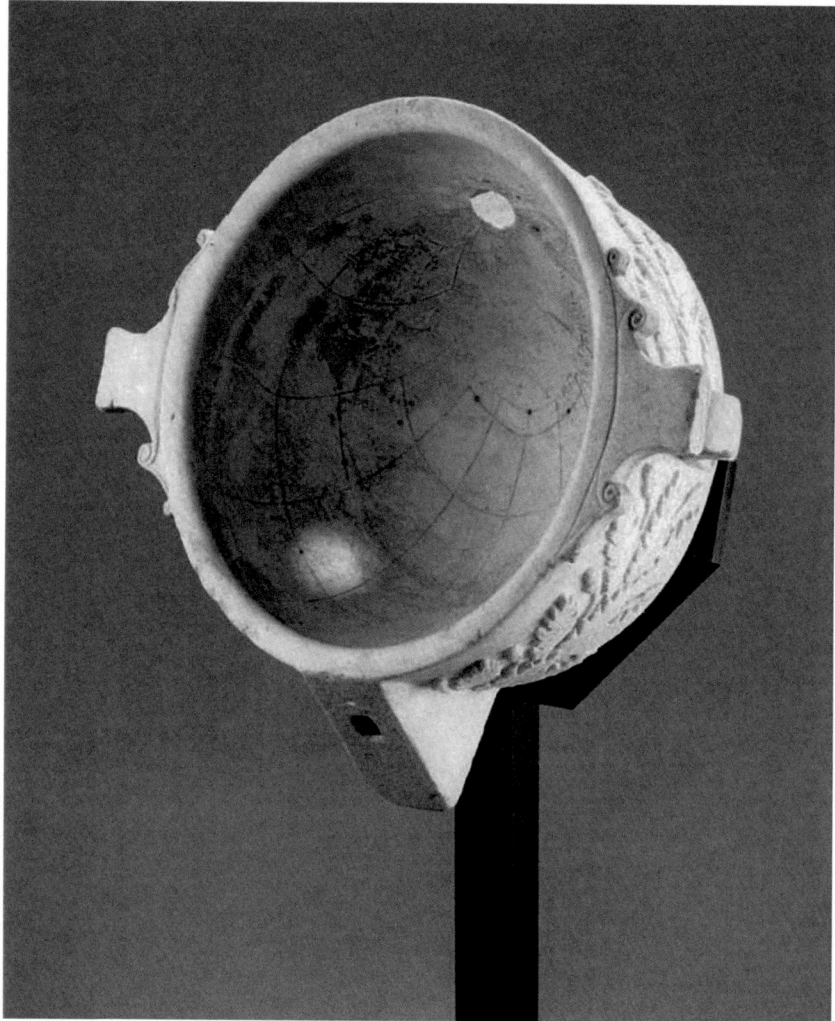

Figure 2.2   A scaphe dial with eyelet (Louvre Museum, 1st–2nd century AD). The spot of light indicates the time and the seasons in this hemispherical dial. (Photo: RMN/H.Lewandowski)

150 meters long and 75 meters wide. It is worth mentioning in passing that the Egyptians themselves never used obelisks as sundials.

No work on gnomonics has survived from antiquity. Only in the *De Archi-tectura* of Vitruvius, a Roman architect of the 1st century BC, do we find any extensive discussion of different types of sundials.

Figure 2.3   A Graeco-Roman dial (Timgad, Algeria). Originally, the style stood at the intersection of the hour lines. The tip of the shadow indicated the temporal hours and the dates of the seasons. (Photo: N. Marquet)

Sundials in the Middle Ages

In the Middle Ages there appeared, mostly on religious buildings, a particular kind of sundial: the canonical dial (Figure 2.4). These were not necessarily meant to be true sundials, but rather indicators of the times of prayer and religious offices. They bear no numbers and usually consist of a semicircle divided into 6, 8 or 12 equal sectors. By convention, when the shadow of the pointer lay along one of the lines, it was time for a religious observance.

A decisive step

The civilization of the Arabs, which inherited some of the astronomical knowledge of the ancient Greeks, developed spherical trigonometry, and introduced a major

Figure 2.4 A canonical sundial (Coulgens, Charente). This dial is on a pillar of the 12th-century church tower. The semi-circular face is divided into six equal sectors. When the shadow of the style, at right angles to the face, fell upon one of the lines, it was time for one of the daily prayers. (Photo: S. Grégori)

improvement in sundials. Instead of using horizontal or vertical gnomons, the Arabs placed the indicator parallel to the Earth's axis of rotation, i.e. pointing at the Celestial Pole. A pointer orientated in this way is known as a polar style. This system meant that hours of a consistent length of 60 minutes could be used throughout the year, and moreover, the whole of the shadow indicated the hour rather than just its tip. In other words, although the length of the shadow changed with the passing seasons, the shadow pointed in the same direction at the same time all year round. The vertical Arab-Islamic sundial shown in Figure 2.5 was designed to determine the hours of Islamic prayers.

The rise of the sundial
In the West, in spite of the development of clock-making from the 13th century onwards, sundials became more and more common. Most of them were of the vertical kind, on the walls of churches, castles and the houses of the rich. They were designed to show the dates of the seasons, sunrise and sunset times, etc. Some sundials were portable, using the variations in the altitude of the Sun during the day to tell the time. Others showed only solar noon, and were used partly to serve as a standard for regulating clocks (Figure 2.6). This type of sundial, known as a

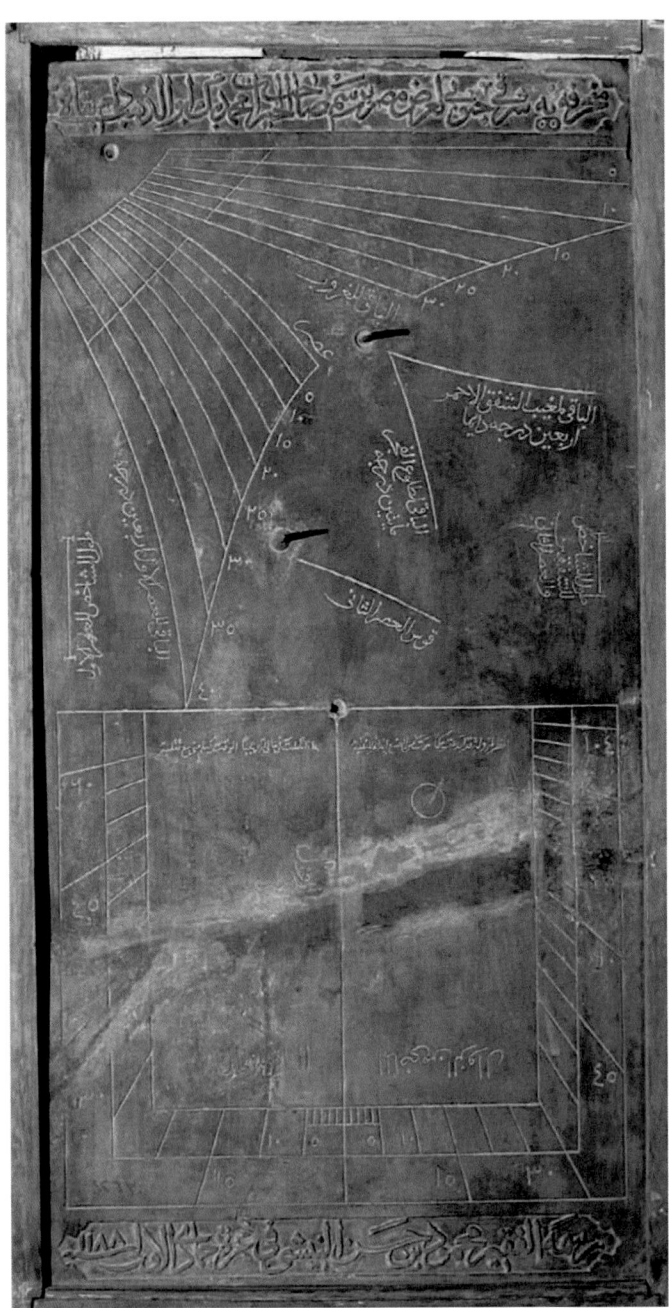

Figure 2.5 An Arab-Islamic sundial in the Museum of Islamic Art in Cairo, Egypt. This magnificent vertical sundial was designed to determine the hours of Islamic prayers. As well as solar time, it indicated the prayers of the *'asr*, the number of hours before sunset and the amount of time before morning and evening twilight. It is signed and dated: "Designed by Mahmoud ibn Hassan El Noushoufi in 1188 H at the beginning of the month of Jumaada 1st" (July 10 1774). (Photo: A. du Boistesselin)

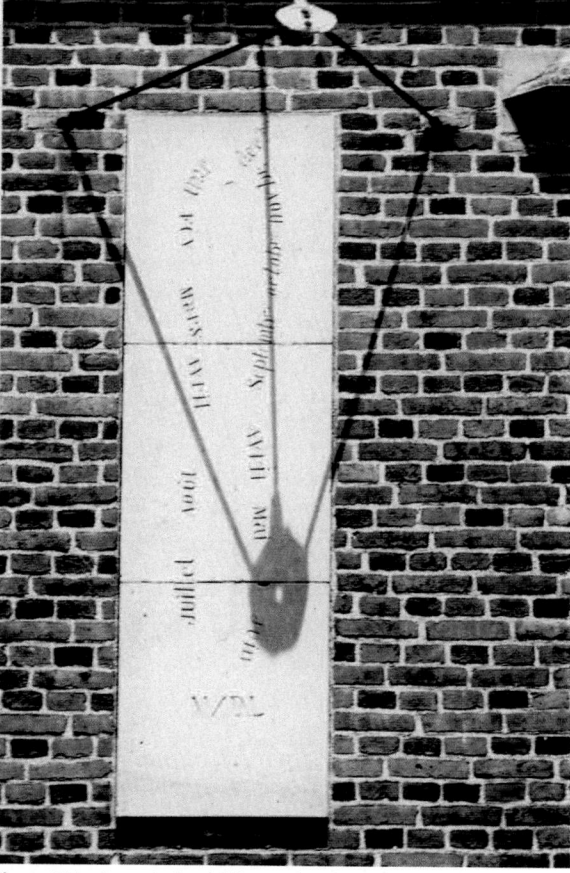

Figure 2.6 The frontispiece of *Gnomonics* (18th century) by Bedos de Celles. This engraving shows the use of *méridiennes* to regulate the time of clocks and watches at solar midday.

Figure 2.7 A vertical *méridienne* showing mean time on the town hall at Aumale (Seine-Maritime). This type of *méridienne* shows mean noon, i.e. true noon with a correction for the equation of time. When the light passing through the eyelet falls upon the figure-of-eight curve, local mean noon can be read as a function of the date.

meridian dial, was sometimes found even inside buildings, examples being the *méridiennes* of the St Sulpice Church in Paris, and of the Paris Observatory.

During the 16th and 17th centuries, ever more ingenious and original dials were constructed: for example, analemmatic types with movable gnomons, reflection dials using a mirror to project a patch of light, and so on. From the 18th century onwards, mean time was indicated on certain dials by introducing the equation of time as a curve shaped like a figure of eight (Figure 2.7). Clocks could now be regulated on the basis of a uniform time (mean solar time), the true solar time of conventional sundials having been non-uniform.

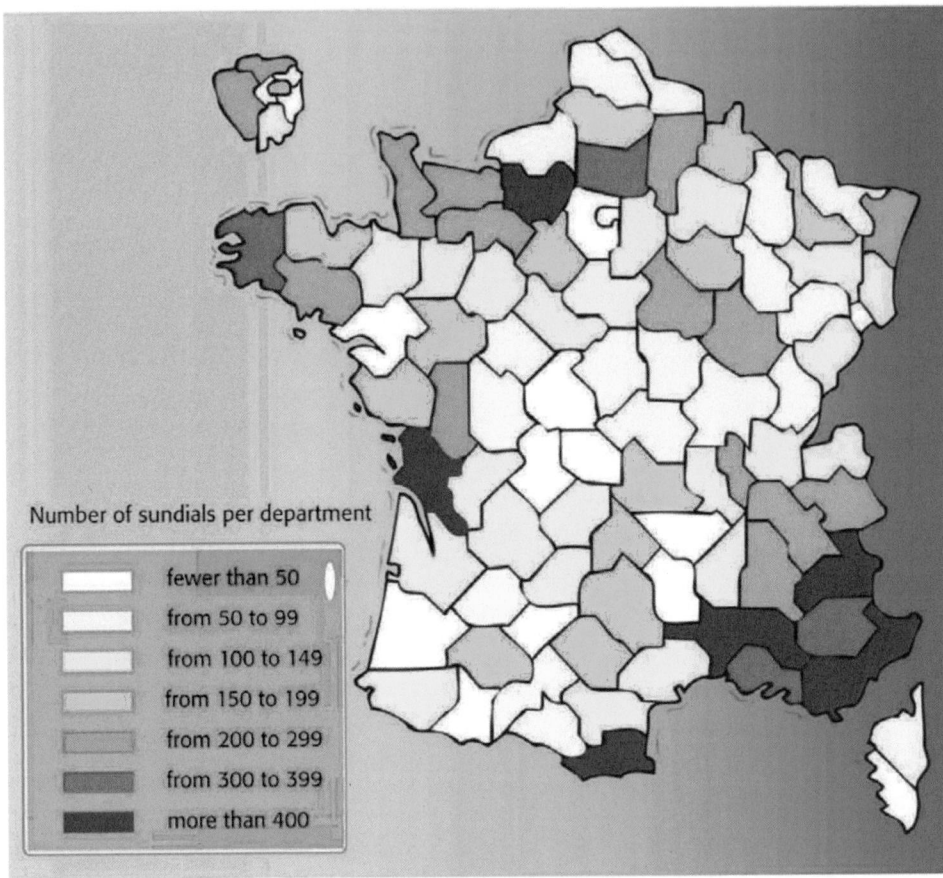

Figure 2.8 A map showing the distribution of sundials across France.

Constructors of sundials, and especially of the more traditional dials, were specialist artisans who calculated and marked out the faces of the instruments, either by painting or sculpting them. To aid their calculations they used published tables, indicating the angles between hour lines as a function of the orientation of the wall and the latitude.

Today, sundials are still made all over the world. Although they do not nowadays play as important a role as in times gone by, they remain objects of interest. They are often adorned with mottoes evoking the passage of time (see Appendix H, page 167). To date, nearly 23 000 sundials have been listed in France (Figure 2.8), in places as varied as churches, castles, public buildings, schools, gardens and many homes.

Measuring the circumference of the Earth with a sundial

In the third century BC, the Greek astronomer Eratosthenes measured the circumference of the Earth, and thereby deduced its radius. His experiment was recounted much later (in the second or third century AD) by Cleomedes.

It was known that, near Syene (modern-day Aswan) in southern Egypt, there was a well, down which could be observed a reflection of the Sun at the summer solstice at noon. In other words, the Sun passed through the zenith above the well, Syene being located close to the Tropic of Cancer.

Now further north, in Alexandria, there stood a pillar which cast a shadow on the ground: it would be a simple matter to measure the length of this shadow and determine the Sun's altitude, from which one could calculate the angular distance between the two towns. Eratosthenes knew the distance between Alexandria and Syene (about 800 km) from the work of the bematists, who were employed to pace out and record distances. He could therefore deduce a value for the Earth's circumference. This method presupposes that Alexandria and Syene lie on the same meridian, as Cleomedes believed, but this is not the case. Their meridians are 3° apart. The important thing here, though, is not the accuracy of the measurement, but the simplicity of the method.

A modern version of the experiment might proceed as follows. The latitude of Alexandria (point A) is 30°. Let us site the well at Syene (point S) on the same meridian as Alexandria, and at latitude 23° 26'. On the day of the summer solstice, the Sun is at the zenith above the well. A gnomon planted in the ground has no shadow. At the same time, in Alexandria, a gnomon of length a casts a shadow ℓ. In order to measure angle α (the angle between the vertical of the gnomon and the Sun), we use $\ell/a = \tan \alpha$, whence we can deduce α. Because the Sun is very far away, its rays are parallel, such that angle α is equal to β, which is the difference in latitude between Alexandria and Syene, a difference of 6° 34'. We know that the distance on the Earth s surface AS $= R\beta\pi/180°$, where R is the radius of the Earth. If we take the distance AS to be 750 km, then $R = 6544$ km (the correct value is 6378 km).

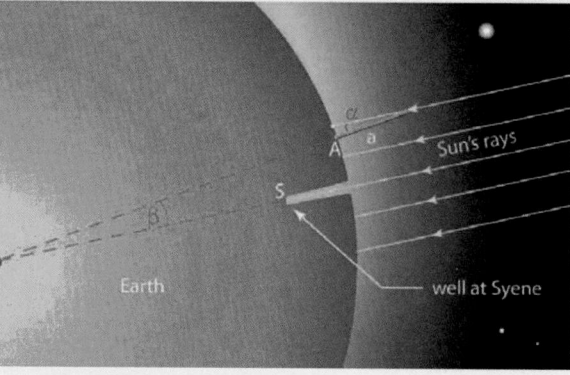

2.2   A brief history of time

Temporal hours

In antiquity, knowing the time was of secondary importance. The rhythm of life was that of the rising and setting of the Sun as the seasons passed, and for most of the time activity was based around such simple markers. However, the rich cities of ancient Greece and Rome had their sundials, indicating the hours: temporal (i.e. unequal) hours. Then, the day was divided into 12 equal parts, no matter what the season. Now we know that the Sun is not above the horizon for an equal length of time in both summer and winter. In France for example, there are about 16 hours of daylight in summer and about 8 in winter. Obviously, if we divide 16 daylight hours by 12 to obtain the length of a temporal hour in summer, we will not arrive at the same result if we divide 8 hours by 12 in winter. In the first case, we obtain a temporal hour of 1 h 20 m, while in the second case the result is 40 minutes. Then, if we perform the calculation at the equinoxes, when the day length is 12 hours, we arrive at a value for a temporal hour of 60 minutes. If we read in some ancient text that an event occurred at, for example, the sixth hour, this does not refer to 6 o'clock in the morning or in the evening, but rather solar noon.

Equal hours

So the Ancients used their variable hours, a practice which continued until the late Middle Ages. Only at the equinoxes did an hour last for 60 minutes. From the thirteenth century onwards, mechanical clocks gradually came into use. It soon became apparent that constructing clocks that showed variable hours was not a practical proposition, more especially since the length of temporal hours also depended upon one's latitude. It was therefore decided that clocks would show equal, 60-minute hours all the year round.

In the days of the French *Convention*, decimal hours, corresponding to 144 modern-day minutes, were instituted. A day had 10 hours, an hour 100 minutes, and a minute 100 seconds. However, this revolutionary time régime only lasted for about sixteen months.

Starting points

The point at which the day is deemed to start has also varied through the centuries. Babylonian hours were reckoned from sunrise onwards (Figure 2.9), and so-called Italic hours from sunset, with variations in other regions. Since the sixteenth century, astronomers have counted the hours from midnight as their starting point.

From local time to Universal Time

Until the beginning of the nineteenth century, every town kept its own solar time. Then, as clocks became ever more accurate, clockmakers were obliged to set their clocks by the Sun, for which task they made use of meridian sundials (*méridiennes*) which indicated only noon (Figure 2.6). However, this solar time was non-uniform,

Why is the day divided into 24 hours?

The idea of dividing the day into 24 hours originated in Egypt. To understand how this came about, we need to look briefly at the Egyptian calendar. The Egyptian year consisted of 365 days, divided into 12 months of 30 days each ($= 360$ days), to which were added 5 extra days, known as epagomenal days. The months were divided into three 'weeks' of ten days each.

In about 2000 BC, Egyptian priests were faced with the problem of determining the hours for night-time prayer, and the idea evolved of using the stars for this purpose. Let us suppose that a certain star A is seen to rise just as the dawn begins to break, and then the star disappears as the dawn glow overcomes it. Let us then declare that this phenomenon, known as a heliacal rising, indicates the end of the night. Star A is considered to be the star marking the last night hour. At dawn the next morning, another brief sighting of star A marks the end of the night, but, as the days pass, it becomes increasingly easier to observe the star in a (darker) sky. It becomes obvious that, after a certain number of days, star A can no longer be used to signal the end of the night. However, there are other stars which can take on this role. The question is: for how long is it necessary to wait to replace A with another star? The Egyptians worked these night hours into their calendar; just as the month was divided into ten-day 'weeks', so the night was marked by timekeeper stars. A star was chosen as the signal for the end of the night for a ten-day period, and then another star for the next ten days, and so on. With 365 days in the year, there were 36 of these 'decans', plus the five epagomenal days.

We might therefore assume that, at any one time, there will be 18 decans above the horizon. But we must remember that it does not become totally dark immediately the Sun sets, and that it is broad daylight even before the Sun has risen; moreover, days and nights are not all of equal length. This means that, because of the unequal lengths of the nights and the duration of twilight, only 12 decans will be observable at their rising. So, the night was divided into 12 hours, as was the day. It is worth noting that it was the decimal division of the month that led to the duodecimal division of the day.

and a correction (the equation of time) was needed to arrive at a uniform time (mean solar time).

As communications, and especially railways, became more highly developed during the nineteenth century, the idea of using the same time all over France became a necessity; in 1891 it was decreed that the time used throughout France would be the mean time of Paris. Some time before this, at an international conference in Washington in 1884, the decision had been taken to divide the world into 24 time zones, and to select a unique meridian from which longitudes were to be reckoned. The meridian passing though the Royal Greenwich Observatory was chosen. Only in 1911 did France begin to tell its time by reference to the Greenwich meridian. Time as defined with reference to the Greenwich meridian is known as

Figure 2.9   A vertical sundial in the courtyard of the Hôtel des Invalides in Paris. The black, unbroken hour lines indicate true solar time, and the red dotted lines indicate the Babylonian hour (i.e. the time elapsed since sunrise). (Photo: SAF/R.Sagot)

Universal Time (UT), although GMT should not be identified with UT as they differ by 12 hours!

In 1916, in order to save energy, Summer Time, one hour ahead of UT, was introduced. In 1976 in France, Summer Time 2 hours ahead of UT, and Winter Time 1 hour ahead of UT were instituted.

During the twentieth century, there were many changes in timekeeping. The unit of time (i.e. the second) had always been defined with reference to the rotation of the Earth, but it became apparent by the end of the nineteenth century that the Earth was not spinning at a constant rate, and was in fact slowing down. It was realized that this deceleration was caused by the ocean tides. Also, clocks were becoming even more accurate (for example quartz clocks), while it was discovered that the Earth exhibited anomalies in its rotation and was in fact sometimes speeding up a little and sometimes slowing. In 1967, the definition of the second was handed over to the physicists, and our modern, worldwide definition is based upon atomic transitions, and is not directly related to any astronomical phenomena.

As for the time as announced by the 'speaking clock', this is known as Coordinated Universal Time (UTC). This time scale is kept by time laboratories

---

Universal Time

Universal Time (UT) is local (civil) time at Greenwich. In France, it is obtained by subtracting 2 h from the clock time when Summer Time is in force, and 1 h when Winter Time is in force. What is Greenwich Mean Time? A sundial measures the hour angle of the Sun, otherwise known as true solar time. When a sundial indicates noon, the hour angle is zero, or 0 h true solar time. Taking into account the equation of time, we obtain mean solar time; when a sundial indicates mean noon, the hour angle of the mean Sun is zero, again 0 h, in mean solar time. For a long time, hours were reckoned in twelves: twelve in the morning and twelve in the afternoon, and indeed we still say, for example, "it's 3 o'clock in the afternoon". However, midnight was seen as the start of the day, and the time when the date changed. When the 24-hour clock came into use, midnight was associated with 0 h. Mean time had to be abandoned, as its starting point was noon, in favor of civil time: mean time plus 12 hours. Imagine a sundial located exactly on the Greenwich Meridian (longitude 0°), indicating solar noon. We subtract the equation of time +12 h and obtain Universal Time. In other words, Universal Time is Greenwich Mean Time + 12 h. So we see that, when we say that it is 12 h GMT, it is not the same as saying that it is 12 h UT.

---

around the world, and is determined using highly precise atomic clocks. The International Bureau of Weights and Measures makes use of data from the timing laboratories to provide the international standard UTC which is accurate to approximately a nanosecond (one billionth of a second) per day.

UTC is the time transmitted by standard radio stations that broadcast time signals, and it may be obtained from Global Positioning System (GPS) satellites. UTC is equivalent to the civil time in Iceland, Liberia, Morocco, Senegal, Ghana, Mali, Mauritania, and several other countries. During the winter months, UTC is also the civil time scale for the United Kingdom and Ireland.

Time zones around the world are expressed as positive or negative offsets from UTC. Local time is UTC plus the time zone offset for that location, plus another offset (typically +1 h) for daylight saving time, if that is in effect.

2.3 The general principle of the sundial

A sundial consists of a surface called the **dial table** (or **dial plate**) upon which is marked a set of hour lines, and a **style**, the shadow of which falls upon the dial table. The table may be horizontal, vertical or inclined at an intermediate angle. As for the style, it may be set into the dial table, or be parallel to it, and it may point at the pole of the sky or be perpendicular to the table.

Tracing out a dial face involves projecting the circles of the Celestial Sphere onto a plane surface. Let us imagine a transparent sphere with center O, of arbitrary radius, upon which are marked the Equator, the two Tropics, and merid-

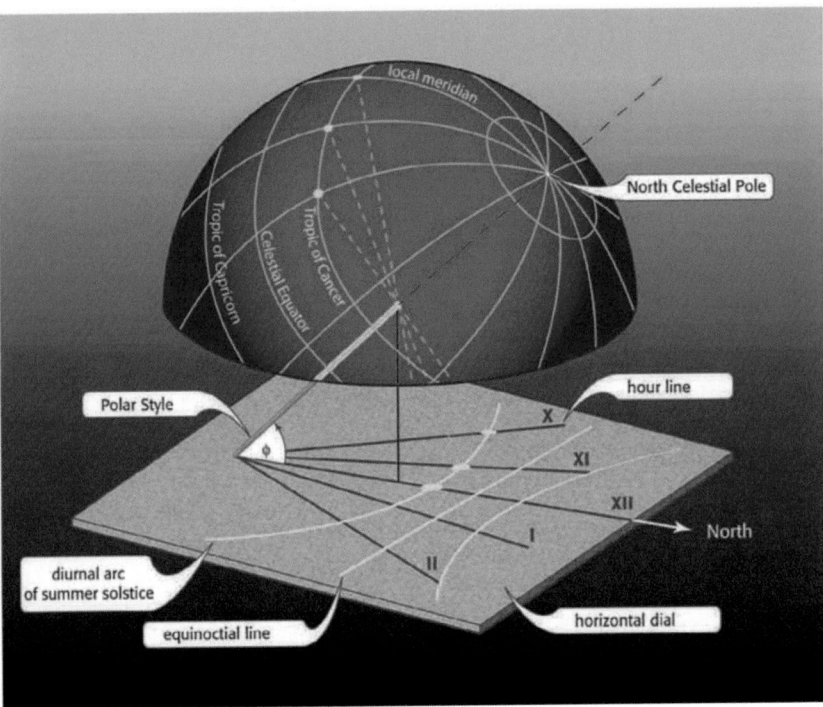

Figure 2.10a The principle of the sundial. (a) The meridians are projected onto the horizontal sundial from the hour lines, and the parallels from the diurnal arcs.

ians 15° apart. Then let us incline this sphere relative to the surface such that its axis of rotation points towards the North Celestial Pole (Figure 2.10a). If we illuminate the sphere with a lamp representing the Sun, situated in the plane of the Equator of the sphere, it will become apparent that the projections of the meridians upon the ground will be in the form of straight lines—the hour lines of the dial. The Equator will also be projected as a straight line, known as the equinoctial line, and the two Tropics will be projected in the form of hyperbolae: the **diurnal arcs**.

On the sphere the meridians are 15° apart, because the Sun travels through 360° in 24 hours; but it will be seen that their projections upon the ground are no longer 15° apart. Depending upon whether the meridians are projected onto a horizontal, vertical or inclined surface, the angles between the hour lines will be different (Figure 2.10b).

These hour lines converge (as projected onto the ground) at a point corresponding to the intersection of the axis of the sphere with the ground, although it is possible that this point will not exist, for example if the plane of projection is parallel to the axis of rotation.

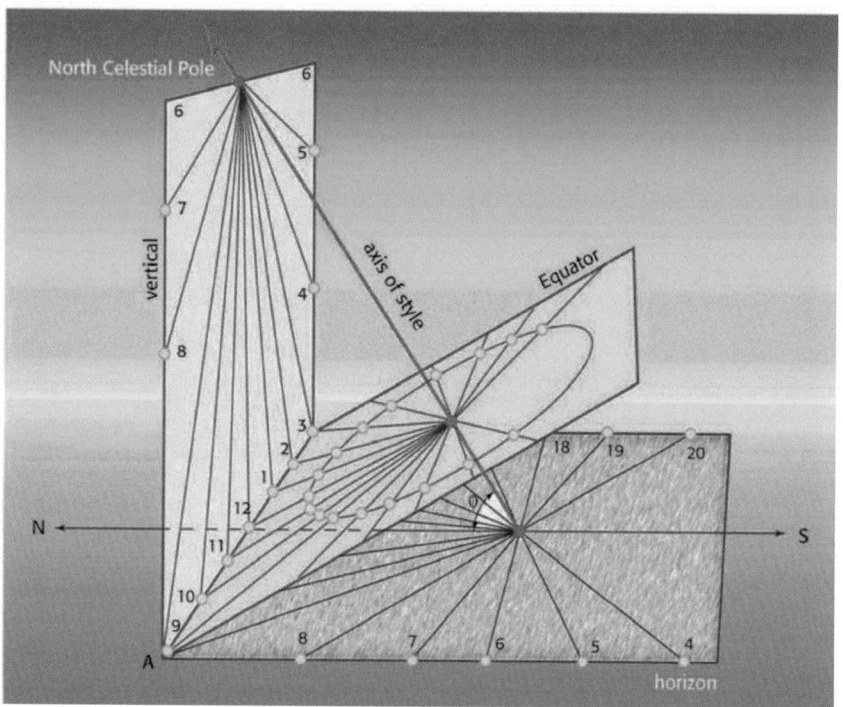

Figure 2.10b The principle of the sundial. (b) The meridians 15° apart in the equatorial plane can be projected onto variably inclined planes, in the form of straight lines known as hour lines.

If we now remove the sphere, leaving only its polar axis, we find that we have made a horizontal sundial with its hour lines, diurnal arcs and an axis which is called the **polar style**.

With the aforementioned arrangement, it will be appreciated that, if the Sun lies in the plane of a meridian (a great circle through the poles), the angular distance of the dial from the Equator can be varied without altering the direction of the shadow cast by the polar style (Figure 2.11). In more technical terms, this means that the direction of the shadow is independent of the declination of the Sun, in the case of a sundial with a polar style.

What degree of accuracy should we expect from a sundial? If it is correctly oriented and figured, the best we can achieve is a reading correct to one minute. Accuracy beyond this is illusory—sundials are not intended to show the time to the very second, and moreover, there are other factors which work against accurate readings: these include penumbral effects, variation in the equation of time, and refraction.

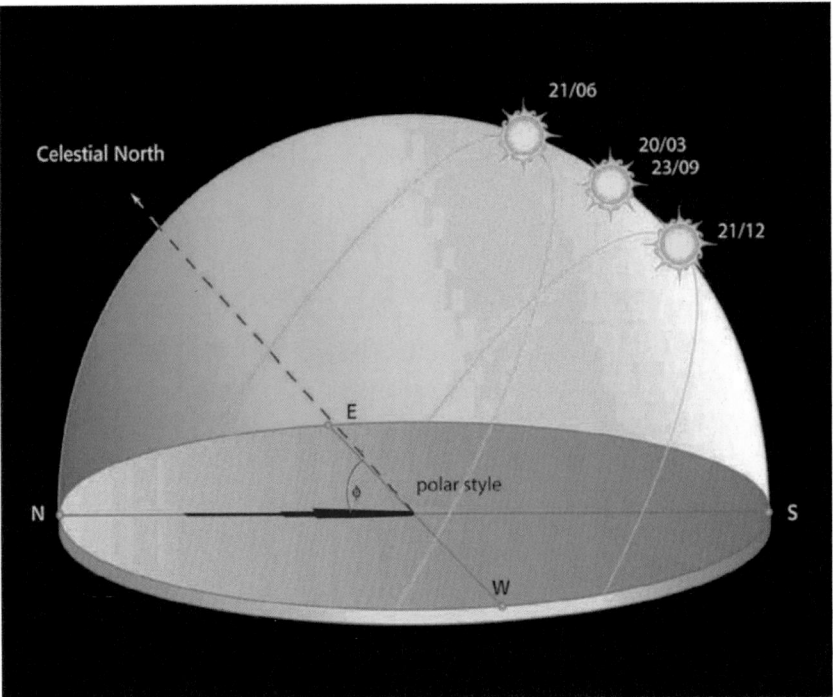

Figure 2.11 When the Sun is in the plane of the meridian, the direction of the shadow of the polar style does not vary.

2.4   The conversion of solar time to clock time

Solar time and local time

All the sundials described in this book show true solar time. The first thing we tend to do after installing our sundial is to compare the time it shows with the time on the clock: and we immediately realize that they are different. There are three reasons for this.

The first involves the Sun itself. In Chapter 1 we saw that its motion is not uniform: it speeds up and slows down relative to a Sun which would move uniformly and return to the same point every 24 hours. It is therefore necessary to correct solar time by a first quantity known as the equation of time. There is a table at the end of this book giving its value for every day of the year. Remember always that the equation of time can have a positive or a negative value (Figure 2.12). If we subtract the equation of time from the solar time as shown by the sundial, we obtain mean local solar time. However, a second correction is now needed, as mean time depends upon the location of the observer.

Figure 2.12 A *méridienne* by Chavin at Serres (Hautes-Alpes). The figure-of-eight curve represents the equation of time. (Photo: S. Grégori)

It is important to bear in mind that a sundial situated, for example, in Strasbourg cannot show the same time as one in Brest, since they do not lie upon the same meridian.

Now, at the beginning of the nineteenth century, every town in France kept its own time, but it became necessary for practical reasons to refer to a single time, and for a while this was Paris time. Then, in 1911, France adopted the time system based on the Greenwich meridian. Places not on this zero-longitude meridian need to adjust their solar times according to their own longitudes, adding if the place lies to the west of the Greenwich meridian and subtracting if to the east.

So we have now introduced two corrections to our sundial time, for the equation of time and for our longitude. The time we arrive at is known as Universal Time (UT). There is, however, yet another correction to be made.

In 1916, France introduced Summer Time, by adding one hour to Universal Time, and in 1976, Winter Time (1 hour ahead of UT) and double Summer Time (2 hours ahead of UT).

In summary, to convert solar time as observed on a sundial in France to local clock time, we must

  subtract the equation of time;
  correct for local longitude;
  add 1 hour if Winter Time is in force and 2 hours if Summer Time is in force.

  Local Clock Time = Solar Time − equation of time + longitude + 1 h or 2 h

Conversion: example

Let us suppose that, on 21 June, we are in Strasbourg, and we read a sundial time of 13 h. We know that the correction for the longitude of Strasbourg is −31 minutes. In the table for the equation of time, we see that, on this date, the value for the equation of time is −1 min 44 s. As we are on Summer Time in June, we have to proceed as follows:

$$13\,h - 31\,min - (-1\,min\ 44\,s) + 2\,h = 14\,h\ 30\,m\ 44\,s$$

(which we can round up to 14 h 31 min).

On 1 December in Strasbourg, with the sundial time still 13 h, we are on Winter Time. In the table for the equation of time, we see that, on this date, the value for the equation of time is +11 min. So the time on the clock is:

$$13\,h - 31\,min - 11\,min + 1\,h = 13\,h\ 18\,min$$

We now move to Nantes (longitude = +6 min 12 s). On 21 June, when the sundial reads 13 h, the clock time is approximately 15 h 08 min. On 1 December, with the sundial still reading 13 h, the clock time is 13 h 55 min.

All these conversions may seem a little cumbersome, so tables are often published incorporating all three: the equation of time, longitude correction and the one-hour correction for Universal Time. This means that we can read the correction directly for Winter Time, and have only to add one hour during Summer Time (in France).

2.5 Finding the local meridian

Establishing the local meridian amounts to determining the line running due north and south through the observer's position. If we intend to construct a sundial, this is the first operation we have to perform. Using a compass is definitely not to be recommended, because compasses indicate magnetic north, which is not the same as geographical north. The difference between the two may amount to several degrees, and changes over time. Moreover, the presence of metallic material may cause considerable deviation of the compass needle. The best solution is to use the

Sun, either by observing successive positions of the shadow of a gnomon, or by determining the moment when the Sun crosses the local meridian.

'Equal shadows' method

This method is easy to set up (for example in a school playground), although it is strictly accurate only around the solstices. It gives good results, if great accuracy is not being sought.

On a perfectly level surface, we set up a gnomon (a rod) 50 cm long, and check that it is indeed vertical with the aid of a plumb line. Around the gnomon we draw five circles on the ground, of radii 25 cm, 50 cm, 100 cm, 150 cm, and 200 cm. During the day, the tip of the shadow of the gnomon will pass though each circle at two points. We carefully mark each of these points on each circle, taking into account penumbral effects. At the end of the day, we join the corresponding points on each circle with string. The line passing though the mid-points of the strings and the gnomon indicates the local meridian (Figure 2.13).

'Meridian passage' method

In Europe and North America, the Sun lies exactly due south as it crosses the local meridian. At this instant, the shadow of a gnomon or a plumb line indicates due north. By knowing the exact local time of the Sun's meridian passage, we can calculate the line of the meridian by using the shadow. The time of this passage

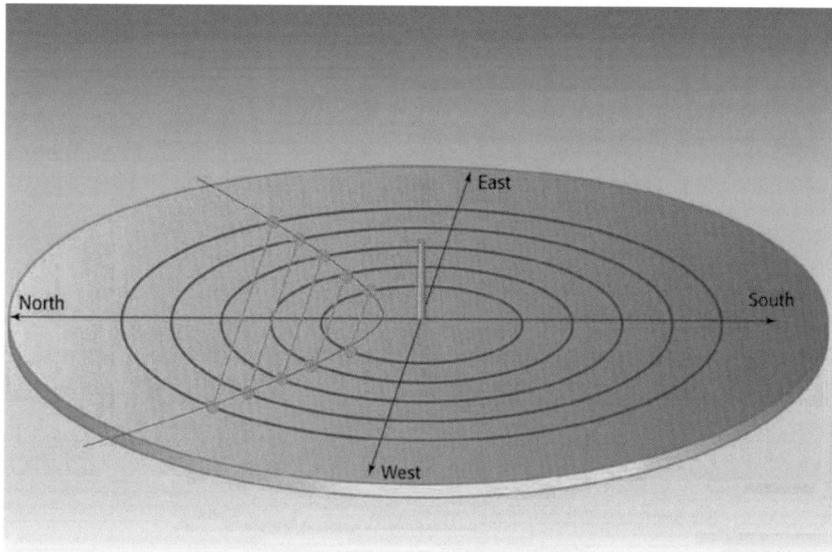

Figure 2.13   The equal shadows method. The red curve shows the path of the shadow of the tip of the gnomon during the day.

varies each day, and moreover depends on the observer's longitude. We can find information on this in astronomical ephemerides or, more simply, use the table of the equation of time given in Appendix C, page 146. We proceed as follows.

Knowing our longitude λ (expressed in hours, minutes and seconds) and the value E of the equation of time, we apply the relationship:

$$\text{Time of passage} = 12\,\text{h} - E + \lambda + (1\,\text{h or } 2\,\text{h})$$

We must be attentive to the sign (+, −) of the equation of time and longitude, or we lay ourselves open to considerable errors! Also, it is necessary to know whether we are on Winter Time or Summer Time. In France, for the former, we add one hour, and for the latter, two hours. The result we obtain is local clock time, and we can verify the accuracy of our clock by using the speaking clock service.

Example 1. Let us suppose that we are at a place at longitude −0 h 22 m 54 s. We are finding the meridian line for 20 May (i.e. in Summer Time). We read from the table of the equation of time that, on this date, E = +3 min 30 s. The time of the Sun's meridian passage is therefore:

$$\text{Time of passage} = 12\,\text{h} - 3\,\text{min } 30\,\text{s} - 22\,\text{min } 54\,\text{s} + 2\,\text{h} = 13\,\text{h } 33\,\text{min } 36\,\text{s}$$

Example 2. Let us now suppose that we are at a place at longitude +0 h 17 m 58 s. We are finding the meridian line for 10 February (i.e. in Winter Time). We read from the table of the equation of time that, on this date, E = −14 min 14 s. The time of the Sun's meridian passage is therefore:

$$\text{Time of passage} = 12\,\text{h} - (-14\,\text{min } 14\,\text{s}) + 17\,\text{min } 58\,\text{s} + 1\,\text{h} = 13\,\text{h } 32\,\text{min } 12\,\text{s}$$

Knowing the time of the Sun's meridian passage, we set up a tripod on a level surface. From the tripod we suspend a plumb line. We can counteract the oscillations of the plumb line by lowering the plumb-bob (but not the string!) into a bowl of water. At the exact moment of meridian passage, we mark the position of the shadow cast by the plumb line on the ground. It is advisable to repeat this operation at least twice, in order to ensure accuracy. The line thus marked can then be drawn out to a greater length.

In spite of its simplicity, this method has two drawbacks: the first is that, if a cloud passes by at the moment of the meridian passage, we have to try again the next day, or the next, or the next ...! Secondly, and this is more to do with the accuracy of the procedure, the azimuth of the Sun changes rapidly in the summer around noon, such that one small error in timing considerably alters the direction of the 'north-south' line marked. Around the summer solstice, the Sun moves approximately one minute of arc in two seconds, whereas at the winter solstice, the angle is only 30 seconds of arc!

Finally, if great accuracy is desired, it is essential to refer to the moment of meridian passage as given in astronomical ephemerides since, strictly speaking, this can vary by a few seconds for the same day from one year to another (Figure 2.14).

Figure 2.14 The meridian passage method. As the Sun crosses the meridian, the shadow of a plumb-line shows the local meridian on the ground.

A sundial for the blind

An extract from *Récréations Mathématiques et Physiques* (Vol. III), by Jacques Ozanam, Paris 1778:

"Now here is a singular paradox. We shall demonstrate that it is possible to install, at the Hôpital des Quinze-Vingts, for the usage of the blind persons therein dwelling, a sundial which will permit them to know the hour by means of touch. To this end, let us take a glass globe of eighteen inches in diameter, and full of water; its focus will lie at nine inches from its surface and the heat produced at this focus will be considerable enough to be felt by the hand upon which it will fall. Moreover, it is easy to see that this focus will absolutely follow the course of the Sun, since it will be always diametrically opposed thereto.

Let this globe be surrounded by a portion of a concentric sphere, at a distance of nine inches from its surface, consisting only of the two Tropics, with the Equator, and the two meridians or colures. Let this instrument be placed in the sunlight in a correct position, that is, with its axis parallel to that of the Earth.

Let each of the Tropics and the Equator be divided into twenty-four equal parts, and let the corresponding parts be connected by a small bar representing a portion of the hour circle, between the two Tropics: by this means we have all the hour circles, represented in such a way that a blind person will be able to tell them, starting from that representing noon, which can be easily designated in some special wise.

Therefore, when a blind person wishes to know the time from this dial, he will begin by placing his hand upon the meridian, and will count the hour circles using the bars representing them. When he arrives at the bar where the solar focus is situated, he will know it by its heat. Therefore, he will learn by this artifice, how many hours have passed since noon, or how many hours remain before the same.

It will be easy to divide each interval between the principal bars marking the hours by other, smaller bars for the halves and quarters. And thus the problem is resolved."

Further information

- Calculating the Sun's local meridian passage with astronomical ephemerides or the Internet: see Appendix G, page 165.

3 The gnomon

The simplest sundial consists of a stick set vertically in the ground. This is known as a gnomon, from the Greek γνωμων, 'an indicator'. It can also be used as an astronomical instrument.

3.1 The first gnomons

Tradition has it that the gnomon was invented by the Greek Anaximander (sixth century BC), though we know from Herodotus (fifth century BC) that the Greeks drew their knowledge of this instrument from the Babylonians. It is probable that the gnomon was also known in China and India, so that it is not really possible to attribute its invention to any particular civilization. What is certain is that, early on, it was recognized that, if a stick set in the ground cast a shadow, this shadow would have apparently 'infinite' length at sunrise. It would become shorter during the morning, and be at its shortest when the Sun was highest in the sky. The shadow would then lengthen once again to become 'infinite' once more at sunset. As well as this everyday phenomenon, it was noticed that the shadow changed in other ways as the seasons passed: at noon in winter, the shadow was longer at noon than in summer. Therefore, since ancient times, the gnomon has been used to determine the dates of the seasons. Careful observation of the path of the tip of the shadow shows that, at the equinoxes, it describes a straight line on the ground. At the winter solstice, this line curves towards the south, while at the summer solstice, it is curved towards the north. Finally, the gnomon is undoubtedly the most ancient of all astronomical instruments. It was commonly used by ancient geographers to find local latitude. (See fact box on page 48.)

In this chapter we will discuss the gnomon not as a sundial proper, since it is very inaccurate compared with other types of sundial, but as an astronomical instrument with which we can both find the local meridian and determine our latitude.

3.2 Determining local latitude

Theory
Since the altitude of the Pole Star above the northern horizon can be used approximately to indicate the observer's latitude, we might use this star to find this coordinate; but the Pole Star does not lie exactly at the polar point of the sky,

Latitude in antiquity

Here are two examples of how latitude was expressed in antiquity.

In the first century BC, the Roman architect Vitruvius wrote the famous *De Architectura*, dedicated to Augustus. In Book IX of this work, he writes at length about sundials, explaining that the way in which they are marked out depends upon the latitude of their location. To define latitude, for example that of Rome, Vitruvius wrote:

> "At the moment of the equinox, the Sun, situated in the Ram or the Scales, causes a shadow which is 8/9 the length of the gnomon at the latitude of Rome."

What, according to Vitruvius, is the latitude of Rome?

The shadow measured at the equinoxes (when the Sun is 'situated in the Ram or the Scales') corresponds to a declination of zero. Let us call the length of the gnomon 1, and the length of its shadow ℓ. Now, $1/\ell = \tan(90° - \varphi)$, whence $\ell = \tan \varphi$. Since $\ell = 8/9$, we deduce that $\varphi = 41° \, 38'$. So, to express the latitude of a place, authors in ancient times gave the 'ratio of the equinoctial shadow of the gnomon'.

Another, more detailed example comes from the *Almagest*, the most complete work on ancient astronomy. It was written in Greek by the greatest astronomer of antiquity, Ptolemy, in the second century AD. In Chapter VI of Book II, Ptolemy gave a 'table of shadows'. This involved parameters for defining the position of a place on Earth. For the island of Rhodes, he gave the following values:

$$14.5 \text{ hours} \quad 36° \quad 12° \, 55' \quad 43° \, 36' \quad 103° \, 18'$$

The first term represents the duration of the longest day; the second, the local latitude; the third, the length of the shadow of a gnomon of length 60 at the summer solstice (expressed in degrees and minutes, as was the custom in ancient times); the fourth, the length of the shadow of a gnomon of length 60 at the equinoxes; and finally, the fifth, the length of the shadow of a gnomon of length 60 at the winter solstice. As for the obliquity of the ecliptic, the contemporary value was 23° 51'. These values can be easily verified. The longest day is obviously that of the summer solstice; we have seen that the length of the day was equal to $(2H_0)/15$, with $\cos H_0 = -\tan \varphi \tan \delta$. Taking $\varphi = 36°$ and $\delta = +23° \, 51'$, we obtain 14 h 30 min. As for the values for the length of the shadow, they represent the quantities $60 \tan(\varphi - \delta)$, $60 \tan \varphi$, and $60 \tan(\varphi + \delta)$.

where the Earth's axis of rotation meets the Celestial Sphere. It is at present about 0° 44' away from that point, which means that it describes a small circle around the celestial pole as the Earth turns once upon its axis. It is difficult without referring to accurate tables to know the precise moment when the Pole Star crosses the local meridian.

The best solution consists in using the Sun, calculating its height at culmination (Figure 3.1). When the shadow is shortest, the Sun is culminating on the

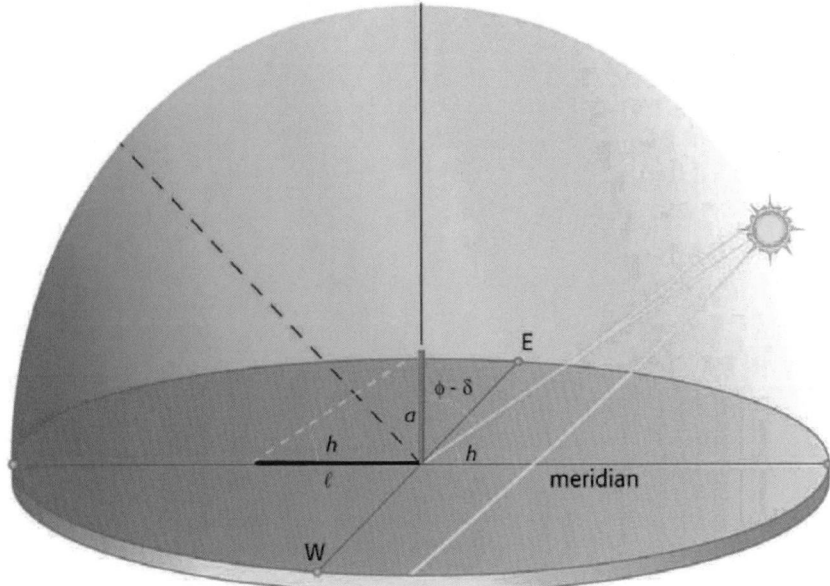

Figure 3.1 The principle of measuring latitude. We can measure the Sun's altitude h at true noon using the length ℓ of the shadow of the gnomon of height a, measured when the Sun culminates: $a/\ell = \tan h$. The local latitude φ is related to h thus: $\varphi = 90° + \delta - h$.

local meridian. It then lies due south. Its altitude h (at true noon) can be calculated using

$$h = 90° - \varphi + \delta$$

where φ is the latitude and δ the Sun's declination. We therefore measure as accurately as possible the length of the shadow of a gnomon at true noon. If ℓ is this length, and a is the height of the gnomon, then $a/\ell = \tan h$. We can now deduce local latitude, since $\varphi = 90° + \delta - h$. At the equinoxes (20 March and 23 September), the declination of the Sun is zero, such that $\varphi = 90° - h$.

Method

- **Materials required:** stick 50 cm long; plumb line.

This procedure allows us simultaneously to determine the local meridian and to measure our latitude. On an open, completely level surface, we set up a vertical stick, 50 cm long. We verify with a plumb line that the gnomon is vertical. We mark the positions of the shadow of the gnomon at various times during the day. The tip of the shadow, at European latitudes, describes a curve which is known as a hyperbola. When the shadow is shortest, the Sun is culminating on the local

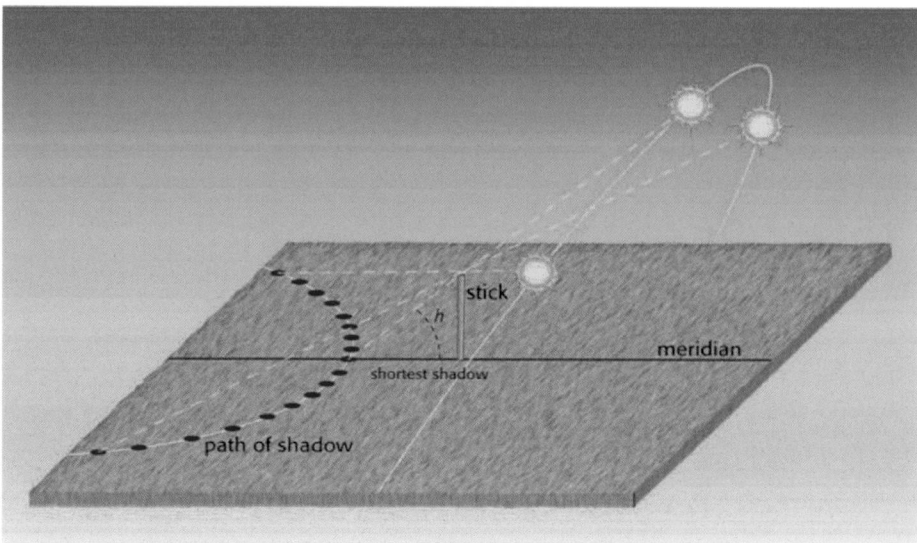

Figure 3.2   To draw the meridian, we observe the position of the shadow of the tip of the gnomon during the day. When the shadow is at its shortest, the Sun is culminating in the south and we can calculate the latitude (as shown in Figure 3.1).

meridian, and lies due south. The local meridian is therefore the straight line running through the foot of the gnomon and the point marking the shortest shadow (Figure 3.2). Marking out this meridian is difficult, since around the time of its culmination, the Sun's altitude shows little variation, and a considerable error could be introduced if we cannot determine the actual shortest shadow. It is therefore necessary to repeat the procedure on several consecutive days. Also, we should avoid doing this experiment around the winter solstice, since at this time the shadow is very long and its extremity is blurred by penumbral effects, which make measurement less easy. The length of the shadow lying along the meridian allows us to determine the latitude.

Example
Let us suppose that, in Boise, Idaho on 23 April ($\delta = +12°\ 40'$), we set a gnomon 50 cm long into the ground, and determine that its shortest shadow is 30 cm long. So,

$$\tan h = (50/30),$$

therefore $h = 59°\ 02'$. We calculate from this that the latitude φ of Boise, Idaho is $\varphi = 90° + \delta - h = 43°\ 38'$. The true value is around $43°\ 34'$: the error is due to the lack of accuracy in the measurement of the shadow, which is actually 29.9 cm long. The accuracy of the observation depends very much on how accurately we deter-

---

Curves described by the tip of the shadow

The tip of the shadow of a stick placed vertically in the (perfectly horizontal) ground describes, in our latitudes, a curve known as a hyperbola. Let us suppose that the foot of the stick is the origin of an orthonormal system, with the x-axis towards the *East* and the y-axis towards the *North*. We can demonstrate that the equation for the hyperbola, in the form $y = f(x)$, is written:

$$y = \frac{-a \sin \varphi \cos \varphi + \sin \delta \sqrt{x^2(\cos^2 \varphi - \sin^2 \delta) + a^2 \cos^2 \delta}}{[\sin^2 \delta - \cos^2 \varphi]}$$

where a is the length of the stick, φ the local latitude and δ the declination of the Sun.

- What form does the hyperbola take at the equinoxes ($\delta = 0°$)?
Answer: After simplification, we obtain y = a (tan φ): the equation for a straight line parallel to the axis of the abscissae.

- What form does the curve take at the North Pole ($\varphi = +90°$)?
Answer: After simplification, we obtain $x^2 + y^2 = a^2/\tan^2 \delta$: the equation for a circle of radius [a/tan δ].

---

mine the exact moment of culmination. Also, the penumbral effect at the tip of the shadow of the gnomon will compromise any measurement of the shadow's length.

In ancient times, this method was long used to calculate local latitudes, and there exist tables giving the length of the shadow for different towns at different times of the year. (See fact box on page 48.)

At the solstices we can also determine the maximum declination of the Sun, the absolute value of which corresponds to the inclination of the ecliptic to the Equator, an angle known as the obliquity of the ecliptic.

Further information:
- Azimuth and hour angle: see Appendix D, page 150.

3.3 The solar calendar

Theory
A solar calendar consists of a gnomon on a level surface upon which the local meridian has been marked. On this meridian are indicated different dates in the year (equinoxes, solstices, festivals, etc.). In order to achieve this, we determine where the tip of the gnomon's shadow crosses the meridian line on a given date for

declination δ. When the Sun culminates in the south, its altitude h can be calculated by using $h = 90° - \varphi + \delta$.

During the year, the Sun's declination δ varies, as does, therefore, the length of the shadow. The values of δ are given by tables (see Appendix C, page 148). Obviously, we must know the local latitude φ. The length ℓ of the shadow is then $\ell = a/\tan h$.

Method

- **Materials required:** stick; plumb line; 'eyelet'.

A solar calendar can be created, for example in a school playground, even by very young children. Assume that we have already had them mark out a north-south line on the ground (see Chapter 2, page 43). Having firmly set up a vertical gnomon of height a, we now wish to determine on what date the tip of the shadow crosses a mark made on the meridian line. We can add a refinement to the system by placing an 'eyelet' at the top of the gnomon, for example a metal disc with a 2-cm hole at its centre. This gives a clearer indication, because the spot of light projected though this hole can now be the indicator, instead of the tip of the shadow: penumbral effects are now eliminated. For this method, the metal disc should be tilted northwards at an angle equal to the local latitude; if we are not seeking great accuracy, an angle of 45° will suffice. Then, we mark along the meridian line those dates we wish to feature on our calendar (Figure 3.3).

Example

Let us suppose that, at Le Mans, France (latitude $\varphi = 48° \, 01'$), we wish to mark on the meridian line of our solar calendar the date of a child's birthday, for example 23 April. For this date, the declination table gives $\delta = +12° \, 40'$. Applying the relationship $h = 90° - \varphi + \delta$, we obtain a value of 54° 39' for the altitude of the Sun.

Let us now calculate the length ℓ of the shadow for this date. Since $a/\ell = \tan h$, it follows that $\ell = a/\tan h$. If the gnomon is 50 cm long, then the length of the shadow will be 35.5 cm. We therefore make a mark on the meridian line at this distance from the gnomon, and on 23 April in any year, we observe that the tip of the shadow (or the spot of light) is indeed at this point. However, we also notice that the same is true on another date: 19 August, because the Sun (except in the case of the solstices) has the same declination twice in the year.

To complete our calendar, we can also mark on the meridian line the dates of the solstices and equinoxes. Using the same parameters as before, we find:

- the length of the shadow at the winter solstice (21 December): $\ell = 149$ cm;
- the length of the shadow at the summer solstice (21 June): $\ell = 22.9$ cm;
- the length of the shadow at the equinoxes (20 March and 23 September): $\ell = 55.6$ cm.

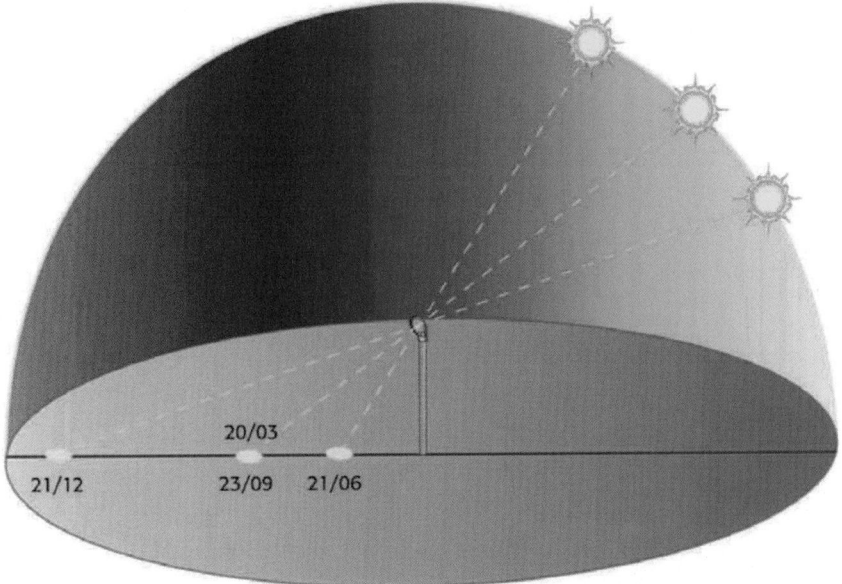

Figure 3.3 The annual variation of the length of the shadow of the gnomon when the Sun culminates can be used to establish a solar calendar. Different dates are indicated on the local meridian passing through the foot of the gnomon.

Another possibility is to indicate the beginning of each month. This is a very good way to show the variation in the Sun's altitude as the seasons pass.

Further information:

- What is the length of the shadow of a gnomon in the course of a day? See Appendix F, page 161.

3.4 A seasonal indicator

Theory

The seasonal indicator is more elaborate than the solar calendar. It requires the use of a complex formula, albeit one that can easily be handled with a computer. A spreadsheet will suffice. On the sundial, we mark out curves joining positions of the tip of the gnomon on special days (solstices, equinoxes and beginnings of months, etc.). This indicator will reveal to us the motions of the Sun throughout the year, and the inequality of the seasons.

Method

- **Materials required:** stick; plumb line; 'eyelet'.

On a level surface, we set up a vertical gnomon of length a. During the day, the tip of the gnomon's shadow will describe curves (hyperbolae) on the surface, except at the equinoxes, when the tip follows a straight line. This simple system can be used to indicate solstices, equinoxes or any other date we choose. We begin by marking out the local meridian, upon which we set up the gnomon. We mark out a pair of orthogonal axes passing through the foot of the gnomon: the y-axis northwards (i.e. along the meridian), and the x-axis eastwards (Figure 3.4). For a given declination δ and a given latitude φ, the x coordinates vary according to the formula:

$$y = \frac{-a \sin\varphi \cos\varphi + \sin\delta\sqrt{x^2(\cos^2\varphi - \sin^2\delta) + a^2\cos^2\delta}}{[\sin^2\delta - \cos^2\varphi]}$$

We obtain the y coordinates along the curve (Table 3.1). In other words, mark out point by point the hyperbola described by the tip of the shadow of the gnomon

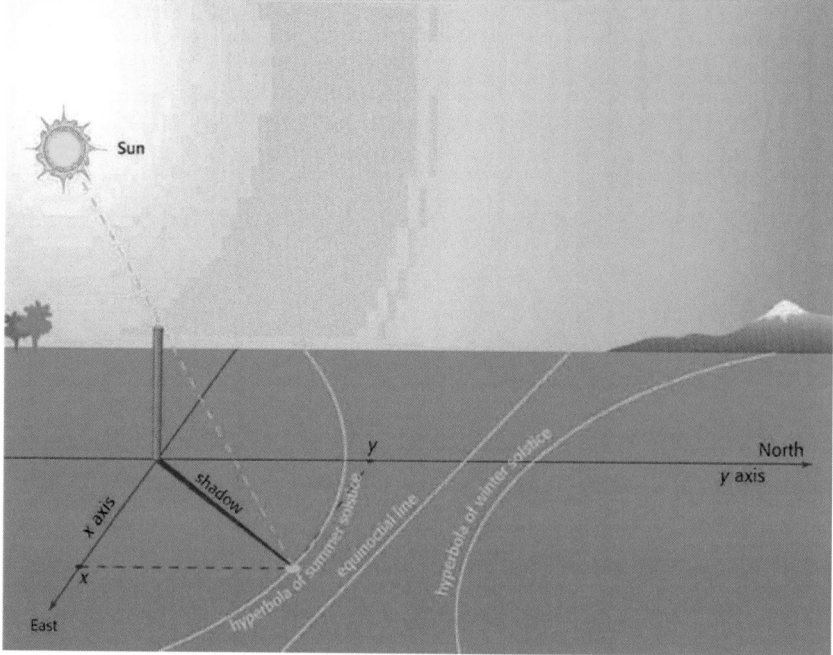

Figure 3.4 Example of a diagram for a seasonal indicator. Three curves are shown: the hyperbolae described by the tip of the shadow of the gnomon at the solstices, and the straight line it describes at the equinoxes.

Table 3.1 Coordinates (in cm) of the points described at latitude $\varphi = 48°$ by the tip of the shadow of a 100-cm gnomon at the summer solstice ($\delta = +23°.44$), the winter solstice ($\delta = -23°.44$) and at the equinoxes ($\delta = 0°$)

	$\delta = +23°.44$	$\delta = -23°.44$	$\delta = 0°$
$x = -200$	$y = -22.54$	$y = 336.07$	$y = 111.06$
$x = -150$	$y = 3.86$	$y = 339.67$	$y = 111.06$
$x = -100$	$y = 25.62$	$y = 317.91$	$y = 111.06$
$x = -50$	$y = 40.39$	$y = 303.14$	$y = 111.06$
$x = 0$	$y = 45.70$	$y = 297.83$	$y = 111.06$
$x = +50$	$y = 40.39$	$y = 303.14$	$y = 111.06$
$x = +100$	$y = 25.62$	$y = 317.91$	$y = 111.06$
$x = +150$	$y = 3.86$	$y = 339.67$	$y = 111.06$
$x = +200$	$y = -22.54$	$y = 336.07$	$y = 111.06$

for a given date. Three curves must be drawn: the summer solstice curve ($\delta = +23°.44$), the winter solstice curve ($\delta = -23°.44$), and the equinox line, which is very easy to determine since the coordinate is constant whatever the value of x: i.e. $y = a \tan \varphi$. Before actually setting up this indicator, it is a good idea to work out on paper what we intend to do. We should first consider the amount of space available on the ground, and from this, determine the length of our gnomon: the constant upon which the size of the diagram will depend. It should be borne in mind that the hyperbolae curve away quite sharply, especially in winter.

Example
Let us take as our example a set of curves derived from a seasonal indicator at latitude 48°, with a gnomon 100 cm long. Table 3.1 shows some of the coordinates of the points on the curves corresponding to the solstitial and equinoctial declinations of the Sun.

Note that there is no point in calculating negative values of x because they are all symmetrical with their positive counterparts!

As in the case of the solar shadow calendar, we can mark out on the dial the hyperbola for a particular value of the Sun's declination, e.g. for the beginning of a month, or a birthday, etc. Another refinement is to fix a ball or 'eyelet' at the top of the gnomon, in order to counteract penumbral effects: these become considerable when the Sun is low down near the horizon. For historical reasons, many sundials

Table 3.2	Declination of the Sun as it enters the various signs of the Zodiac	
Declination	*Dates*	*Signs of Zodiac*
$\delta = -23°.44$	21 December	Capricorn
$\delta = -20°.15$	20 January and 22 November	Aquarius and Sagittarius
$\delta = -11°.47$	19 February and 23 October	Pisces and Scorpio
$\delta = 0°$	20 March and 23 September	Aries and Libra
$\delta = +11°.47$	20 April and 23 August	Taurus and Virgo
$\delta = +20°.15$	21 May and 23 July	Gemini and Leo
$\delta = +23°.44$	21 June	Cancer

include curves representing notable values for the Sun's declination, as it enters the various signs of the Zodiac. The corresponding dates and declinations are summarized in Table 3.2.

4 Equatorial sundials

The equatorial sundial is an inclined dial. It is very easy to mark out. However, its workings are not always immediately understood. It is certainly the best instrument to teach us about the movement of the Sun in declination. It is called an *equatorial* dial because its table, graduated on both sides, is set up parallel to the Earth's Equator.

Figure 4.1 The north face of an equatorial sundial (Châteaubernard, Charente). (S.Grégori)

4.1 The principle of the equatorial sundial

The dial

In the case of an equatorial sundial (Figure 4.1), the style, casting its shadow on the dial table, is parallel to the Earth's axis of rotation. It is aligned with the local meridian and makes an angle with the horizontal plane equal to the local latitude φ (Figure 4.2). The dial table, at right angles to the style, is therefore parallel to the Equator and is at an angle of $(90° - \varphi)$ to the horizontal. The table has hour lines on both surfaces. Why is this? As we saw in Chapter 1, the Sun is above the Celestial Equator in both spring and summer; i.e. its declination is positive. This means that the north-facing surface of the table is then illuminated. In autumn and winter, however, the Sun lies below the Equator and the south-facing surface is illuminated (Figure 4.3). At the equinoxes, the Sun is exactly on the Equator, and both faces are illuminated.

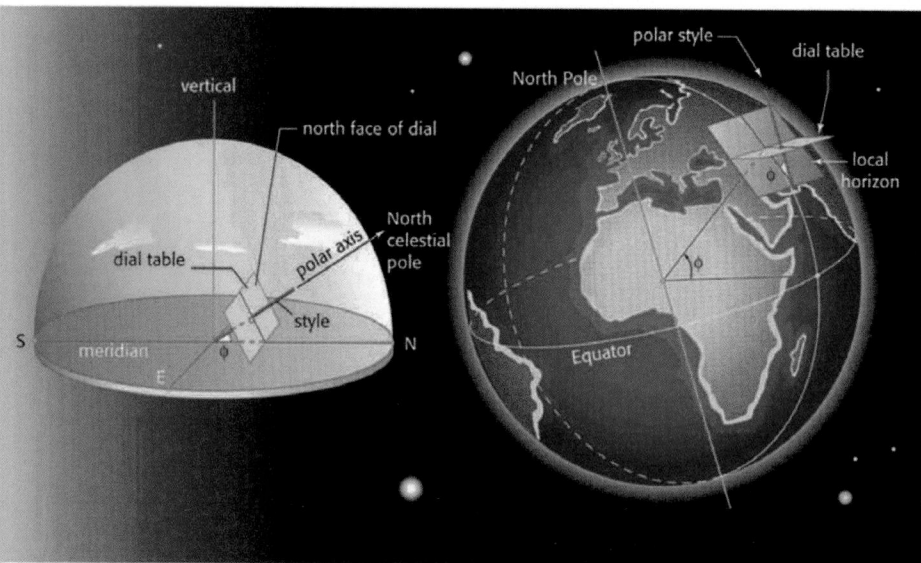

Figure 4.2 The orientation of an equatorial sundial (a) to the local Celestial Sphere and (b) on the Earth. The style is parallel to the Earth's axis of rotation and is situated on the local meridian. The table is at right angles to the style and parallel to the Equator.

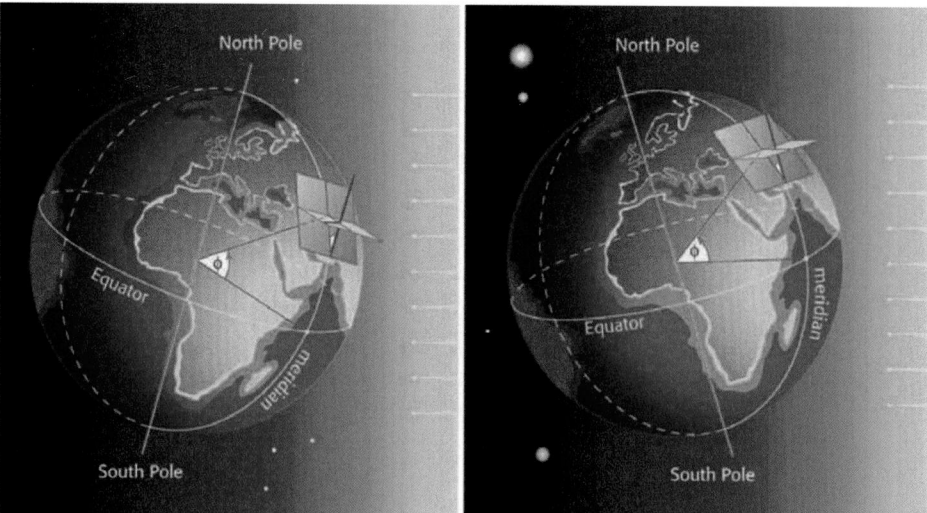

Figure 4.3 The annual evolution of the shadow of the style. In spring and summer (a), the Sun is above the Equator and the north face of the dial is illuminated. In autumn and winter (b), the Sun is below the Equator and the south face of the dial is illuminated.

The shadow of the style

As the day passes, the shadow of the style does not move in the same direction across both faces: as we look at the north-facing side, the shadow travels in a clockwise direction, but on the south-facing side it travels anti-clockwise. Also, on the north-facing side, the shadow of the style is shortest on the day of the summer solstice. After 21 June, the shadow becomes longer, and infinitely long at the autumn equinox. On the south-facing side, the shadow is obviously at its shortest at the winter solstice. If we record the positions of the tip of the shadow throughout the day, we see that they lie on a circle. So it is possible to mark out notable dates by marking out concentric circles.

Hour lines

Since the Sun moves at a rate of $15°$ per hour along the Equator and the table is parallel to the Equator, the angle between each hour line is also $15°$. On the day of the summer solstice, the Sun is $23°\ 26'$ above the Equator. In France, this date corresponds to the earliest sunrise (at 4 h true solar time) and the latest sunset (20 h). The northern face must therefore be graduated with hour lines from 4 h to 20 h, to accommodate the earliest and latest sunlit hours at our latitude (Figure 4.4). Now at the winter solstice, the Sun rises later (8 h) and sets earlier (16 h). These are not the outermost hours marked on the southern face of the dial, however, because it is illuminated until the equinoxes, the days when the Sun rises at 6 h and sets at 18 h. So the southern face is graduated from 6 h to 18 h.

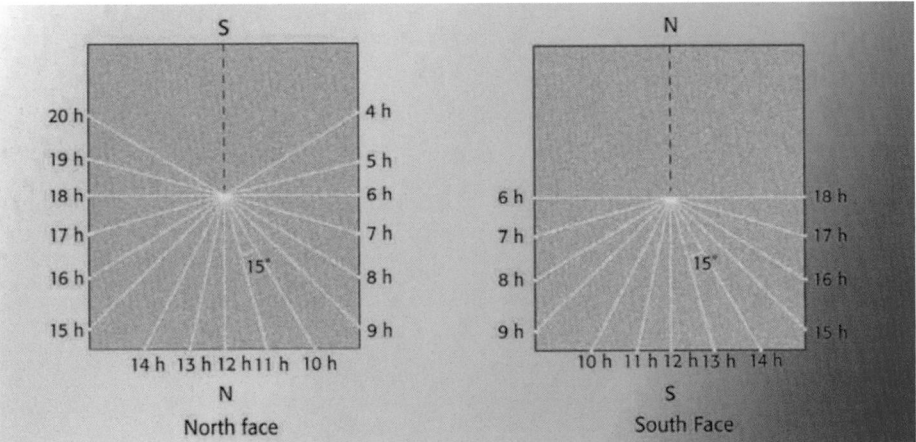

Figure 4.4 Hour lines on an equatorial sundial. The north face is graduated from 4 h to 20 h, and the south face from 6 h to 18 h. The angle between the hour lines is $15°$.

4.2   Marking out the classic equatorial sundial

Method
- **Materials required:** 30-cm by 30-cm wooden board 0.5 cm thick; threaded rod 20 cm long; 2 nuts; 2 washers; pair of compasses; protractor; calculator.

We find the centre of the 30-cm by 30-cm board by drawing two diagonals. On the north face of the board we draw a line perpendicular to the edge, passing through the centre. This is the noon line (Figure 4.5). We use the compasses to draw a circle

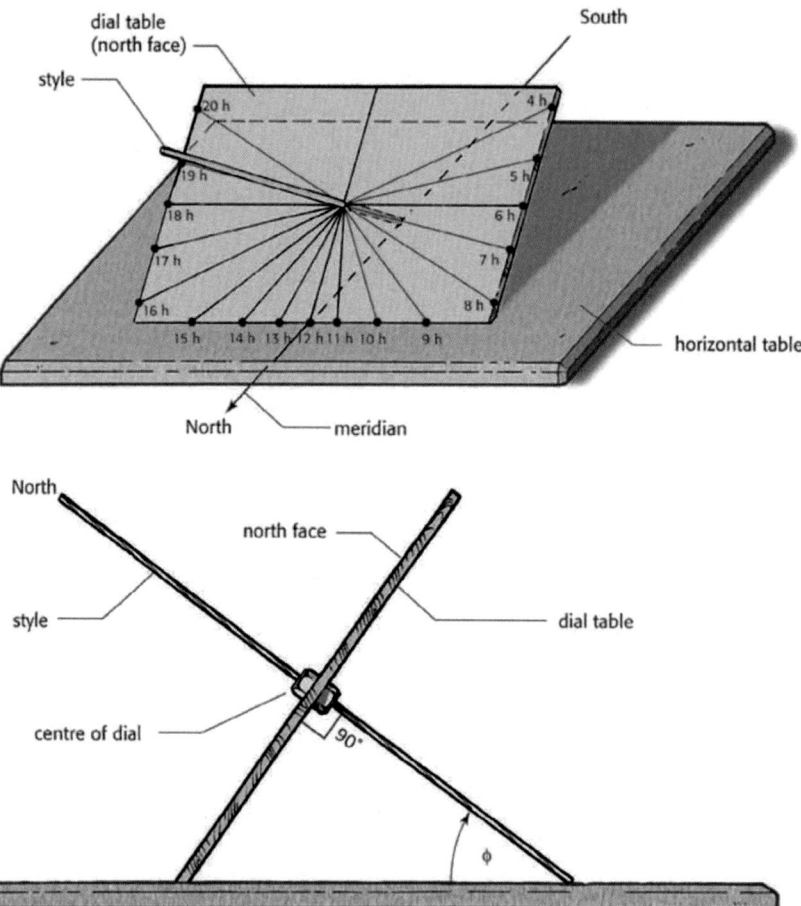

Figure 4.5 The construction of an equatorial sundial.

Let us imagine that we are at the North Pole ($\varphi = +90°$). Here, the dial table is horizontal and its style is vertical. Only the northern face of the dial is functional, and only from the spring equinox to the autumn equinox. At the Equator ($\varphi = 0°$), the equatorial sundial is vertical (like a wall); it still has two working faces, the northern and southern faces being illuminated for 6 months each.

---

of radius 15 cm around the centre, and we use the protractor to divide this into $15°$ sectors, radiating from the centre. In this way, we have marked out eight lines on the left-hand side (13 h to 20 h), and eight lines on the right-hand side (11 h to 4 h). On the south face, we also draw a noon line corresponding to the one on the north face. We then proceed as for the north face, but we only mark our hour lines ($15°$ apart) from 6 h to 18 h. There can be no line beyond the one at right angles to the noon line. Take great care when marking out the lines.

With all the lines marked, we make a hole for the threaded rod through the centre of the board. Having fixed the rod in place with the nuts and the washers, we check with a set square that the style is indeed at right angles to the dial face. Then we set up the sundial. The two noon lines (on the north and south faces) must lie along the local meridian, which we will have already marked out on the ground (or even determined directly with the dial: see below). The rod on the south side must make an angle with the meridian equal to the local latitude φ (Figure 4.5). To ensure that this is so, we can use a protractor, but for even greater accuracy we might make the following calculation: the length of the rod on the south side must equal $15/\tan \varphi$. For example, if $\varphi = 48°$, the length of the rod on the south side is 13.5 cm. If we use a board of dimensions other than 30 cm by 30 cm, the length of the southern rod will be equal to the length of the side squared divided by 2 and further divided by $\tan \varphi$. For example, if our board is 15 cm by 15 cm, the southern rod should be 6.75 cm long for latitude $48°$. The sundial is now ready for use. In spring and summer, the north face will be illuminated and in autumn and winter, the south face.

4.3 Uses of the equatorial sundial

As a calendar

It is possible to find the date by using the equatorial sundial shown in the previous section. The dial table is parallel to the Equator, so the angle between the sunlight grazing the tip of the style and the surface of the sundial is equal to the Sun's declination. Also, the path of the shadow tip travels in a circle as the day passes. This is known as the declination circle. It is easy to demonstrate that the radius of a declination circle is equal to $PK/\tan \delta$ (Figure 4.6). At the equinoxes, this radius is infinite. To show the date on the dial, we first of all have to calculate the length PK

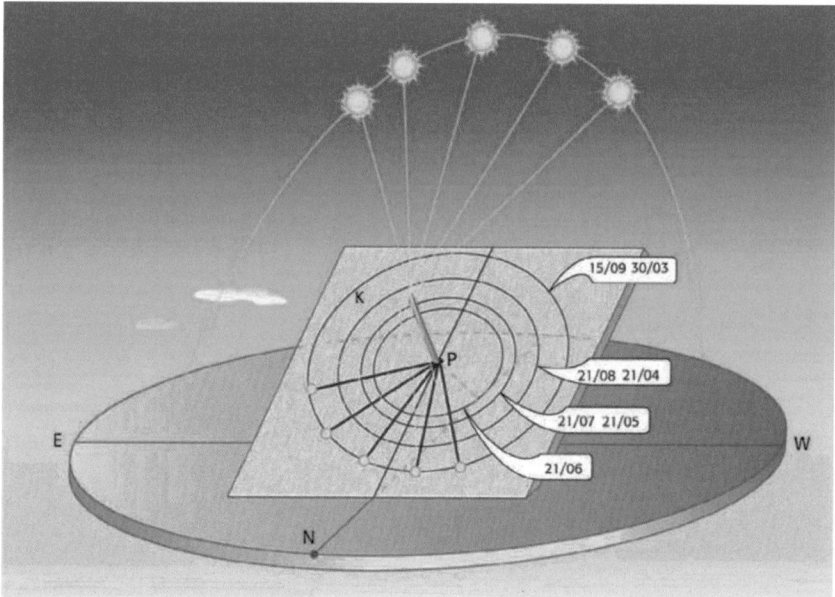

Figure 4.6 The shadow of the tip of the style in the course of a day. On a given date, the shadow of the tip of the style describes a circle of radius PK/tan δ, where δ is the declination of the Sun. To use an equatorial dial as a calendar, we draw several declination circles corresponding to the selected dates.

of the rod on the north face, so that at the summer solstice (δ = 23° 26′), the shadow will follow, for example, a circle of radius 5 cm. This length is 5 × tan(23° 26′), i.e. 2.2 cm. On the southern side, the length of the rod is still 15/tan φ, so the total length of the threaded rod is (5 × tan 23° 26′) + (15/tan φ)+ the thickness of the board (0.5 cm). If φ = 48°, the length is 16.2 cm. All we need to do now is to mark out different declination circles for the different dates we wish to show on the sundial.

　　To indicate a date on the south face, where the rod has no free tip, we place a nut at a predetermined place on the rod. The shadow of the nut will then indicate the date. This solution can also be used on the north side.

Local time
In order to read local time from an equatorial sundial, we need only to slide onto the style a circular card (or piece of wood), with hours marked upon it (Figure 4.7). For use with a 30 cm × 30 cm table, we cut out a circle of card (or wood) of radius 15 cm, with a central hole. We mark out 15° sectors, as for the classic dial, but this time only on the north side. Then we slide it down the rod, and it is superimposed upon the dial table, and can turn around the rod. If we align its noon line with the

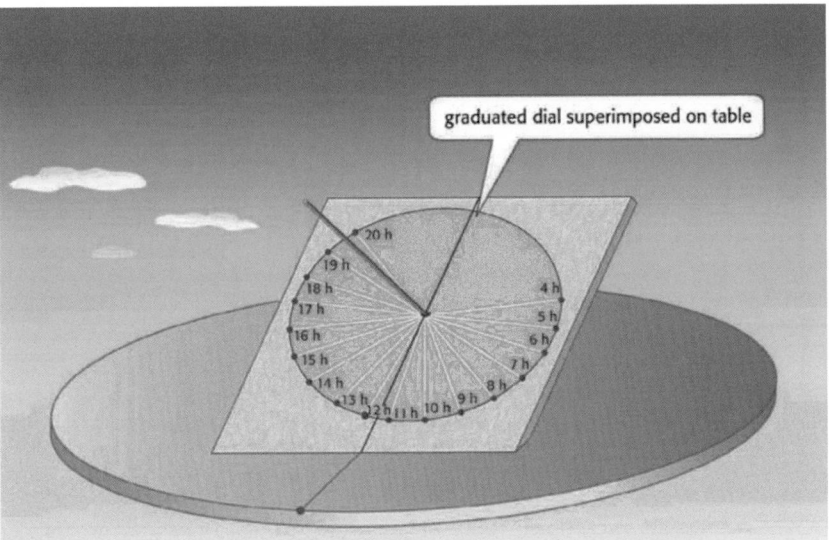

graduated dial superimposed on table

Figure 4.7 The principle of the equatorial dial showing local time. A movable graduated circle is placed over the equatorial dial. The circle is rotated until it shows local time, as shown on a clock.

meridian, we see before us the classic dial; but we can rotate the card around the dial, and it will indicate the time on the clock. For example, if the time is 10 h on the clock, we turn the card until the shadow of the style coincides with the 10 h line on our mobile dial. Thus regulated, the card will tell us the local time all day long. If we leave the card in this position, we will see that in just a few days an error builds up, since the value of the equation of time changes every day.

4.4 Determining the meridian

Theory

The method described here may also be used with a horizontal sundial (see Chapter 5). Assume that we have a correctly calculated and marked equatorial sundial, with its style mounted at the correct angle. We now have to align the dial with the meridian. Let us call the true solar time as shown by the sundial TS, the equation of time E, the local longitude λ and TL the local time as on the clock. All we have to do is calculate the solar time TS as shown by the sundial, when the clock indicates the time TL. This clock will have been set by the speaking clock service. The formula for converting solar to local time is:

$$TL = TS - E + \lambda + (1\,\text{h or }2\,\text{h}).$$

Once we have calculated *TS*, we move the sundial until the shadow of the style indicates the correct time.

Example

Let us imagine an equatorial sundial at longitude -10 min. What is the time on the clock when the dial shows 10 h on 23 April? From the table of the equation of time (see Appendix C, page 146), we find that, on that date, $E = +1$ min 40 s. As Summer Time is in force, we add 2 h. So the clock shows 10 h $-$ 1 min 40 s $-$ 10 min $+$ 2 h, i.e. 11 h 48 min 20 s.

At the predicted time, we align the sundial so that the shadow of the style lies on the 10 h line. This operation can be repeated several times in order to ensure that the meridian has been correctly determined. One advantage of this method is that it is not necessary to have previously marked out the local meridian with the aid of a gnomon or a plumb line. So, once our equatorial sundial is complete, and assuming the Sun actually shines, we can use it at any time during daylight!

4.5   The armillary equatorial sundial

Theory

In ancient times, the armillary sphere was an astronomical instrument used to measure the positions of heavenly bodies. We can modify this sphere, transforming it into an equatorial sundial: instead of reading the solar time on an inclined plane with two faces, we cause the shadow of a style to fall on a band which is a segment of a circle (Figure 4.8). With this arrangement we can tell solar time all the year round, without the complication of having to know which face is illuminated. The only disadvantage of this type of dial is that setting it up and regulating it require some 'DIY' skills.

Method
- **Materials required:** embroidery frame[1] 20 cm in diameter; protractor; threaded rod 20 cm long; small wood-saw.

Embroidery frames found in specialist shops are made of wood and comprise two concentric hoops. Here, we shall use the complete inner hoop. We divide it into 15° sectors, marked from 4 h to 20 h. For this purpose we can lay the hoop on a sheet of paper, measure its interior diameter, determine the centre and then use a protractor to create our sixteen sectors of 15° each.

With the saw, we cut off the section of the hoop between 20 h and 4 h, so that only part of the wooden circle remains (Figure 4.9). We then set up the hoop so that

[1] Translator's note: Sometimes known as a tambour, this is an arrangement of circular hoops over which fabric is stretched while being worked upon.

Figure 4.8
An armillary
sundial
(Cognac,
Charentes).
(S.Grégori)

it makes an angle of $(90° - \varphi)$ with the southern horizon. As in the case of the classic equatorial sundial, the 'noon' point of the hoop must coincide with the meridian line on the ground. Then we set up the threaded rod so that it passes through the exact centre of the arc of the hoop, pointing northwards at an angle to the meridian equal to the local latitude. The rod must be perpendicular to the plane of the hoop.

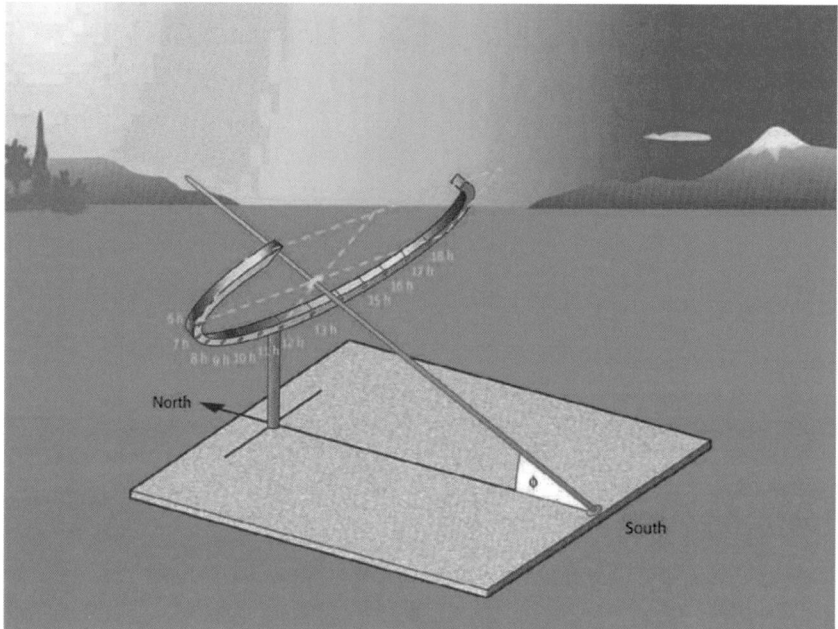

Figure 4.9   The principle of the armillary sundial. The arc upon which we read the time is at an angle to the southern horizon of (90° − φ). The style is at right angles to the plane of the hoop and passes virtually through the centre of the arc.

    With this system, the shadow of the rod falls onto the hoop, parallel to its hour lines, and we can read the time from whichever side we are looking. Obviously, it is possible to graduate the hoop into half-hours too, by dividing it into sectors measuring 7°.5.

5 Horizontal sundials

The horizontal sundial is easy to make. It may be based on an equatorial dial, or created with the help of a little trigonometry and a pocket calculator. It is easily installed in an open location, and works all the year round from sunrise to sunset.

5.1 The principle of the horizontal sundial

The dial

The horizontal sundial comprises a horizontal table with hour lines marked, and a style which points towards the North Celestial Pole and throws a shadow onto the table (Figure 5.1); when the shadow lies upon one of the lines, it indicates that hour in solar time. As always, the dial should be aligned with the local meridian in order

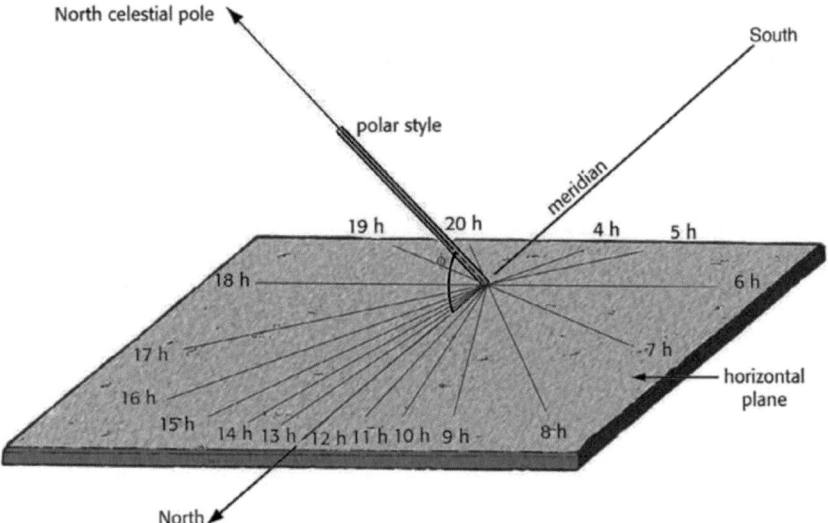

Figure 5.1 The horizontal sundial. On the horizontal table are the hour lines from 4 h to 20 h. The style points towards the North Celestial Pole, and is at an angle φ to the table (the angle of local latitude). The noon line coincides with the meridian.

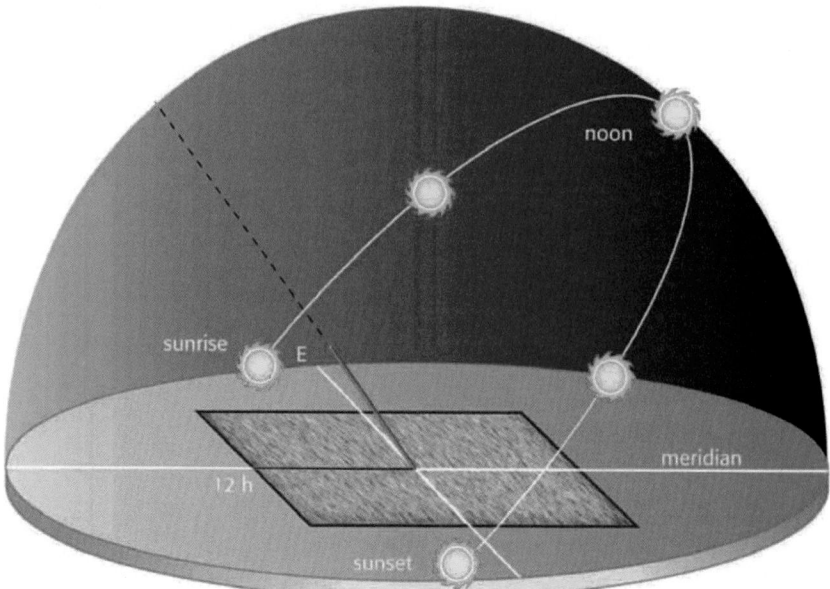

Figure 5.2   The horizontal sundial and the Celestial Sphere. The polar style and the noon line are in the plane of the meridian.

to function properly (Figure 5.2). The style is therefore always in the vertical plane of the meridian.

Hour lines
The hour lines are half-lines passing through the foot of the style. Since the dial works throughout the year, all the lines from 4 h to 20 h, incorporating the extreme times for sunrise and sunset in our latitudes, must be represented (Figure 5.3). The difficulty in marking out these lines lies in the fact that they are at differing angles, and we need to draw them using geometry or trigonometry.

The horizontal sundial at other latitudes

At the North Pole ($\varphi = +90°$), the horizontal sundial becomes an equatorial sundial: the angle between the hour lines is 15° and the style becomes a vertical gnomon. At the Equator ($\varphi = 0°$), the polar style no longer intercepts the dial table, but is parallel to the dial, and the hour lines are parallel straight lines. It therefore becomes a polar sundial (see Chapter 6 for details of this dial).

Figure 5.3 A horizontal sundial at Vairé in the Vendée. The style is in the form of a triangle. (SAF/R.Sagot)

5.2 Marking out, using the equatorial sundial

Theory

The dial face for the horizontal sundial can be drawn by projecting the layout of the equatorial dial onto the horizontal dial table, taking into account the direction of the style (Figure 5.4). It is often advisable to start with this manual procedure rather than try for an accurate drawing based on trigonometry. Having said that, this geometrical method is merely of academic value, and it is preferable to resort to calculation if an accurate, large-scale sundial is to be constructed. Nevertheless, using an equatorial sundial as a projection tool will tell us immediately that an angle of 15° on an inclined plane will no longer be 15° when projected onto the ground. Moreover, it also gives insight into the difficult case of vertical sundials which do not face south, and also that of analemmatic sundials.

Method

- **Materials required:** existing, working equatorial sundial (see Chapter 4); sheet of A3 paper.

First of all, we take an equatorial sundial, here sometimes known as an auxiliary dial, with a polar style and hour lines, 15° apart, to its edge. We place this dial on a

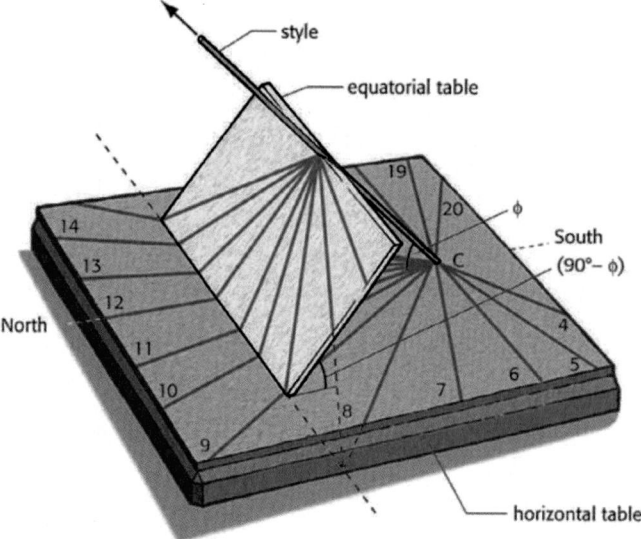

Figure 5.4
Diagram for a horizontal sundial. The hour lines on the dial are obtained by projecting the lines of an equatorial sundial onto the horizontal table.

sheet of A3 paper, width-wise. The A3 sheet represents the table of the horizontal sundial. The style of the equatorial dial meets the paper at point C, where the hour lines of the new sundial converge. At the intersection of an hour line on the equatorial dial and the horizontal dial, we make a mark. From there we draw a line to point C: thereby obtaining a horizontal hour line. The limitations of this method will soon become apparent; hour lines before 9h and after 15h do not reach the horizontal sheet of paper. At best, we can mark out only seven lines!

In traditional books on this subject, the recommendation was to stretch threads across the equatorial dial to prolong the hour lines to meet the horizontal plane. There were also geometrical solutions based upon changing the plane angle, but this could not avoid the fact that certain intersections were well outside the working area.

Nevertheless, this method shows us that the angle between an hour line on the horizontal sundial, for example, the 14h line, does not make an angle of 30° with the noon line as on the equatorial dial.

5.3 Trigonometrical marking

Theory

The diagram for the horizontal dial is a projection of the hour lines of an equatorial dial (Figure 5.5). If we know the local latitude φ, we can easily calculate and mark out the hour lines on a horizontal dial. Angle H' between an hour line and the noon

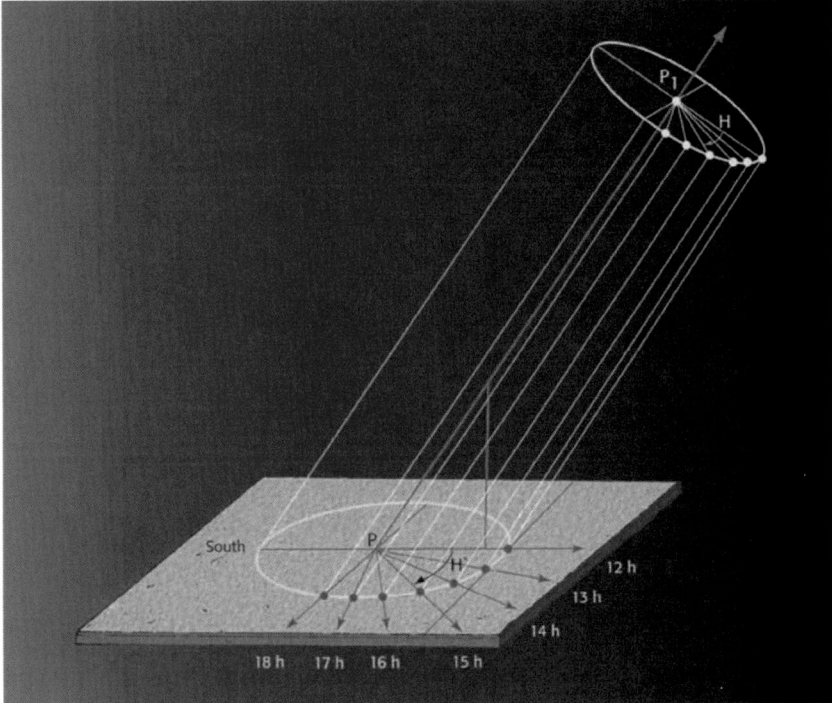

Figure 5.5 The Sun's hour angle H varies by $15°$ per hour in the equatorial plane, while the variation of H' in the horizontal plane depends on the time of day.

line for the horizontal dial is calculated by:

$$\tan H' = \sin \varphi \tan H \tag{1}$$

where H is the hour angle of the Sun ($H = 0°$ for noon, $H = 15°$ for 13 h, $H = 30°$ for 14 h, ..., $H = -15°$ for 11 h, etc.).

Example

In France, the values of H vary between $-120°$ and $+120°$. We therefore calculate H' for these values of H. For example, if $\varphi = 48°$ (at the latitude of Le Mans) and $H = +45°$, then H' will be $36°.62$. If $H = -60°$, $H' = -52°.16$.

Take care if $H = \pm 90°$: then, equation (1) above is invalid. In this case, we use:

$$H = +90° \longrightarrow H' = +90°$$

$$H = -90° \longrightarrow H' = -90°$$

which means that the 18 h and 6 h lines make an angle of $90°$ with the noon line.

Table 5.1 Solar times, hour angle H and angle H' of the hour line associated with the noon line for $\varphi = 48°$

Solar time	Hour angle H	Angle H' of hour line associated with noon line
12 h	0°	0°
13 h and 11 h	±15°	±11°.26
14 h and 10 h	±30°	±23°.22
15 h and 9 h	±45°	±36°.62
16 h and 8 h	±60°	±52°.16
17 h and 7 h	±75°	±70°.17
18 h and 6 h	±90°	±90°
19 h and 5 h	±105°	±109°.83
20 h and 4 h	±120°	±127°.84

H and H' are always of the same sign; morning hours correspond to negative hour angles, and afternoon hours to positive hour angles.

Another important point: H' must be of the same sign as H. Consequently, if, for a positive value of H, our calculation leads to a negative result, we add 180° to that result. Conversely, if a negative value of H leads to a positive result, we subtract 180°. For example, if $H = +105°$ (19 h line), we obtain $H' = -70°.17$; we therefore add 180°, giving $H' = 109°.83$. We can also limit ourselves to calculating H' only for $0° < H < 120°$, since the forenoon angles are symmetrical with the afternoon ones. Table 5.1 shows all the calculations for latitude $\varphi = 48°$.

Having calculated the angles for H', we transfer them to the dial using a protractor and the noon line as a reference. If we wish to draw the half-hour lines, we simply use a 7°.5 spacing.

Method

- **Materials required:** piece of plywood board (24.5 cm × 30 cm); protractor; square wooden board (10 cm × 10 cm, 0.5 cm thick).

How do we now proceed to construct our dial? We take our rectangular piece of wood (24.5 cm × 30 cm), upon which we are going to mark out the dial as shown in Figure 5.6. Leaving a margin 2 cm wide all around the edges, to leave enough room for writing the hour numbers, we draw a rectangle measuring 20.5 cm by 26 cm. The triangular style will be cut from the 0.5-cm thick board (Figure 5.6).

We draw a straight line 10 cm from the left-hand edge of the dial. We then draw another line parallel to the first, 0.5 cm away. The space between these lines will accommodate the width of the style. The two lines correspond to a double noon line.

We then need carefully to position the 6 h–18 h line, at right angles to the noon line. The point where these two lines intersect is where the corner of the style (A) will be inserted. The 6 h–18 h line is drawn 10 cm away from the lower edge of the board, creating enough space for the 4 h and 20 h lines.

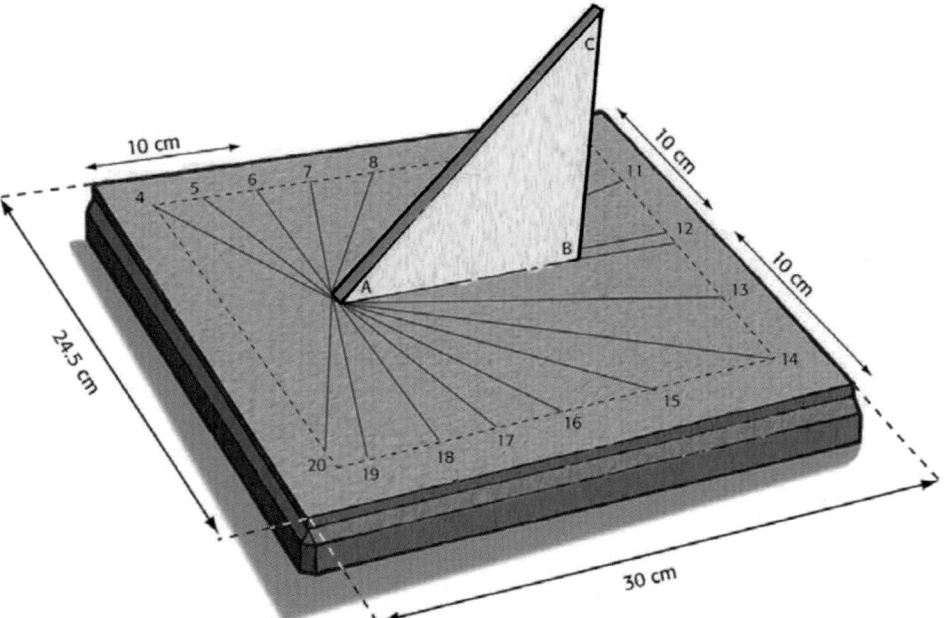

Figure 5.6   A horizontal sundial with a triangular style.

Next, we draw the hour lines at the appropriate angles and the noon lines, to a total of 17 hour lines. Now, care is needed, since the style itself has a certain thickness (0.5 cm): the lines for the hours before 6 h and those for the hours after 18 h do not converge upon the same edge of the style (Figure 5.7). If the style is thick, the shadow of one of its edges will indicate the time before 6 h and after noon, and that of the other the morning hours and those after 18 h. For example, the 19 h line being a prolongation of the 7 h line, it is the shadow cast by western edge of the style, not the eastern, which will indicate 19 h.

As for the triangular style itself, it makes an angle with the noon line equal to the local latitude φ. It can be cut, in the shape of a right-angled triangle, from a piece of board. Let us call the points of this triangle A, B and C: if AC is the hypotenuse and AB the edge attached to the dial, then the height BC of the style $=$ AB tan φ. For example, if $\varphi = 48°$ and AB $= 8$ cm, then BC $= 8.9$ cm.

When the sundial is completed, we align it with the local meridian (see Section 4.4); when the Sun shines on the dial, the shadow of the style will lie along a line, and we read the time from the number marked at the end of that line.

Further information:
• How can we find angles from their tangents? See fact box on page 74.

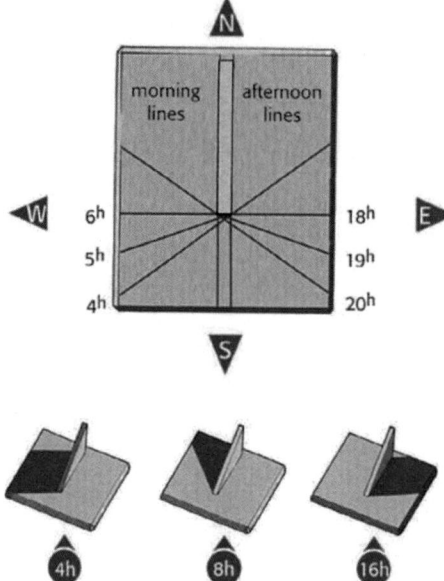

Figure 5.7 Drawing the hour lines for a thick style. The western edge of the triangular style is used to indicate the morning hours after 6 h and the hour after 18 h. These lines originate therefore at the western edge of the style. The eastern edge is used to indicate the hour before 6 h and the afternoon hours until 18 h.

Demonstration of the formula for the horizontal sundial

In the adjoining figure, H is the hour angle of the Sun measured in the equatorial plane, φ is the local latitude, and H' is the angle on the ground between an hour line and the noon line. Now, $BD/BC = \tan H'$. $BD/BA = \tan H$, whence $BD = BA \tan H$. Also, $BC \sin \varphi = BA$, whence $BC = BA/\sin \varphi$. Therefore $\tan H' = BA \tan H/(BA/\sin \varphi)$, i.e. $\tan H' = \sin \varphi \tan H$.

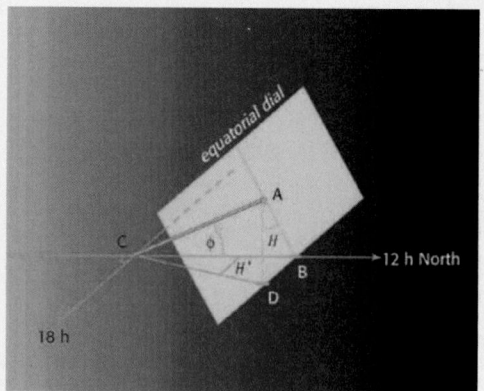

How can we create the angles using their tangents?

Using a protractor may not be a very accurate method. Using trigonometry, it is easy to work out the lengths, almost to the millimeter. For a rectangular dial, let CB be the length of the noon line. At the edge of the dial, we find points consistent with lengths of $(CB \tan H')$. If the lengths project outside the frame, we must take these quantities to a point E $(CE/\tan H')$.

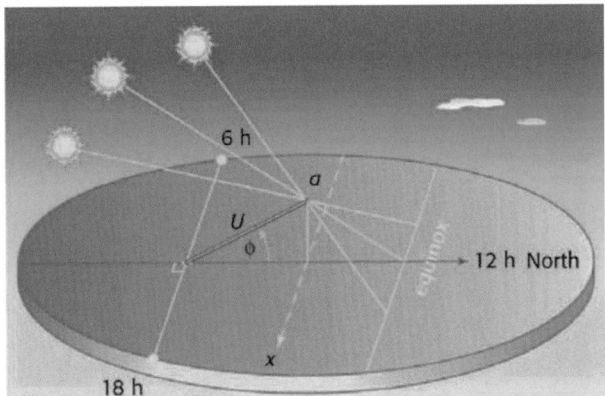

Figure 5.8 The equinoctial. At the equinoxes the shadow of the tip of the style of length U describes a straight line on the dial table at a distance of $U/\cos \varphi$ from the 6 h– 18 h lines.

5.4 Indicating the dates of the equinoxes

While the shadow of the tip of the style of an equatorial sundial describes a circle that is easy to draw, the equivalent curve on the horizontal dial is a hyperbola, which presents much more difficulty. However, at the equinoxes, this hyperbola becomes a straight line, which is much simpler to draw. This line is known as the equinoctial, and is parallel to the 6 h–18 h line, its position depending upon the dimensions of the polar style (Figure 5.8). If we assume a style of length U, then the distance between the equinoctial and the 6 h–18 h line will be $(U/\cos \varphi)$. On the day of the equinox (20 March, 23 September), the tip of the shadow will move along the equinoctial from sunrise to sunset.

We can also dispense with these calculations by following a simple empirical procedure, meticulously marking out the path of the tip of the shadow as the day passes. This method can also be used for other dates, and to mark out the curves for the solstices. So the horizontal sundial may also be used as a calendar (Figure 5.9). Obviously, this method can be somewhat laborious and requires sunshine all day long.

Further information:
- Drawing the hyperbolae: see fact box on page 76.

5.5 Indicating noon for a different location

Theory
It is fairly easy to make a sundial which will indicate true noon in, for example, Cairo, New York, New Delhi, or indeed any place in the world. Let λ be the local

Figure 5.9 A sundial in the park of the town hall in Châteaubernard (Charente). As well as the hour lines, this dial shows the equinoctial and the hyperbolae for winter and summer. (S.Grégori)

Drawing the hyperbolae

In the chapter on the gnomon (Chapter 3), we saw that the equation for the hyperbola, i.e. the curve described by the tip of the shadow of a gnomon, in the form $y = f(x)$, is written:

$$y = \frac{-a \sin \varphi \cos \varphi + \sin \delta \sqrt{x^2(\cos^2 \varphi - \sin^2 \delta) + a^2 \cos^2 \delta}}{[\sin^2 \delta - \cos^2 \varphi]}$$

where a is the length of the gnomon, φ the local latitude and δ the declination of the Sun. Here, the origin of the series of points is at the foot of the gnomon, with the x-axis towards the east and the y-axis towards the north. We can easily use this equation for a horizontal sundial, noting that the tip of the polar style of length U corresponds to the tip of a gnomon of length a. More precisely, the gnomon has a length equal to $(U \sin \varphi)$ such that the origin of the series of points is displaced northwards along the noon line by the quantity $(U \cos \varphi)$. From there, it is easy to locate the points of the hyperbola of the winter and summer solstices, or the hyperbola for any particular date.

Example: If $\varphi = 48°$, $U = 10$ cm, $\delta = +23°.44$, and $a = 7.43$ cm, the series of points is displaced by the quantity 6.7 cm. If $x = 4.12$ cm, then $y = 2.91$ cm. If $x = -23.06$ cm, then $y = -6.69$ cm. If $\delta = -23°.44$, and if $x = -5.93$ cm, then $y = 23.1$ cm.

Having calculated x and y for the same declination, we join up the points between them and obtain a hyperbola, which will be that described by the tip of the shadow on the selected date.

longitude of the sundial, and λ' that of the place for which we wish to indicate noon. Into equation (1) in section 5.3, we put $H = \lambda' - \lambda$. In other words, H represents the difference in longitude between the two places or the two noons. As previously mentioned, we must be careful not to confuse positive and negative signs.

Example
Let us imagine a horizontal sundial at Nice (longitude $\lambda = -7° \; 16'$, latitude $\varphi = 43°$ $42'$. We want it to indicate noon in Cairo, Egypt ($\lambda' = -31° \; 15'$). The difference in longitude between Nice and Cairo is $-31° \; 15' - (-7° \; 16') = -23° \; 59'$, which can be expressed in terms of time (dividing by 15: see Appendix G, page 165: -1 h 35 min 56 s. This means that, when it is noon in Cairo, the time in Nice is 10 h 24 min 04 s. To mark out the noon line for Cairo on the horizontal sundial we therefore use equation (1) in the form: $\tan H' = \sin \varphi \tan(\lambda' - \lambda)$, or in the case of Cairo $H' = -17°.09$. We therefore mark out this angle to the left of the noon line (since H' is negative). If we then proceed to mark out the noon line for New York ($\lambda' = 74°$), we obtain (again for Nice) $H' = 77°.46$, since $H = +81° \; 16'$: when it is true noon in New York, it is 17 h 25 min 04 s in Nice.

5.6   Marking out a horizontal sundial without calculation

Theory
It is quite possible to construct a horizontal sundial without using trigonometrical or geometrical formulae. We proceed entirely empirically, with only two requirements: the presence of the Sun, and a clock!

We begin by determining the local meridian (see section 2.5). At the location where we wish to set up the sundial, we place the style, which must be at an angle with the meridian equal to the latitude. Once the style has been correctly positioned, we synchronize the clock carefully with the speaking clock service. Then, we determine a (solar) time from the local time on the clock. At the time chosen, we mark out the shadow projected by the style on the ground.

Example
Suppose that the local longitude is -19 min, and the latitude is $48°$. Having set up the polar style, on a sunny 23 April, we know that we can find local time TL from solar time TS with:

$$TL = TS - E + \lambda + (1 \text{ h or } 2 \text{ h}),$$

E being the equation of time and λ the local longitude. Since it is 23 April, and Summer Time is in force, we use the formula

$$TL = TS - E + \lambda + 2 \text{ h}.$$

In the table (see Appendix C, page 146) showing the equation of time, we read that, on this date, $E = +1$ min 40 s. Table 5.2 indicates the clock time calculated for each

Table 5.2 Corresponding solar and local times (at latitude 48° N, longitude −19 min) for 23 April

Solar time	Local time
6 h	7 h 39 min 20 s
7 h	8 h 39 min 20 s
8 h	9 h 39 min 20 s
9 h	10 h 39 min 20 s
10 h	11 h 39 min 20 s
11 h	12 h 39 min 20 s
12 h	13 h 39 min 20 s
13 h	14 h 39 min 20 s
14 h	15 h 39 min 20 s
15 h	16 h 39 min 20 s
16 h	17 h 39 min 20 s
17 h	18 h 39 min 20 s
18 h	19 h 39 min 20 s
19 h	20 h 39 min 20 s

hour of solar time. We mark the shadow of the style on the hour for each hour line. If we are fortunate enough to have constant sunshine, we can finish the dial in a day. It is preferable to carry out this operation at a time of year when there are plenty of hours of daylight.

5.7   Babylonian and Italic time

Theory

It is possible to show on a sundial how many hours have elapsed since the Sun last rose (Babylonian hours), or how many hours have elapsed since it last set (Italic hours).

To this end, we construct a rather special horizontal sundial: the style is replaced by a cone. We begin by marking out, in half-hours, a traditional horizontal sundial for the local latitude. We change the numbering of the lines by multiplying by two: the 6 h line becomes 12 h, the 11 h line becomes 22 h, the 14 h line becomes 4 h (24 h + 4 h) etc. At our latitude, we obtain a fan-shaped spread of lines from 8 h to 16 h. Lines to the left of the 0 h line (i.e. on the western side) will indicate Italic hours, while those to the right of the 0 h line (i.e. on the eastern side) will indicate Babylonian hours (Figure 5.10).

Method

- **Materials required:** piece of plywood board (20 cm × 30 cm); protractor; sheet of thick A3 paper.

Making the cone is quite simple, but it must be made from a robust and sufficiently flexible material. On the A3 paper, we mark out a circle of radius R (for example 10 cm). Using the protractor, we draw, from the centre of the circle, a sector of angle α such that $\alpha = (1 - \sin \varphi) \times 360°$, where φ is the local latitude. For example, if $\varphi = 47°$, α will be $96°.7$. We then cut out this sector, and glue together the two edges, thereby obtaining a cone of an opening angle which is double the latitude (i.e. if $\varphi = 47°$, then the opening angle is $94°$). For greater accuracy, it is preferable to attach the cone to a base; for this purpose, we cut out a circle of stiff card, its radius equal to $R \sin \varphi$, and fix this circle to the base of the cone (if $R = 10$ cm and $\varphi = 47°$, the radius is 7.3 cm). Once we have made the cone, we place it on the horizontal sundial (Figure 5.11) so that the apex of the cone is at the point of

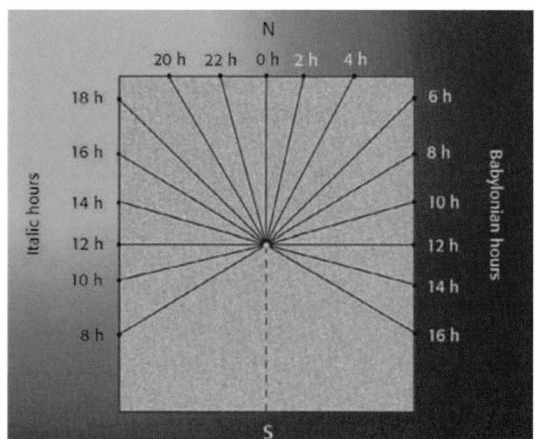

Figure 5.10 Hour lines on a horizontal sundial indicating Italic hours to the west and Babylonian hours to the east.

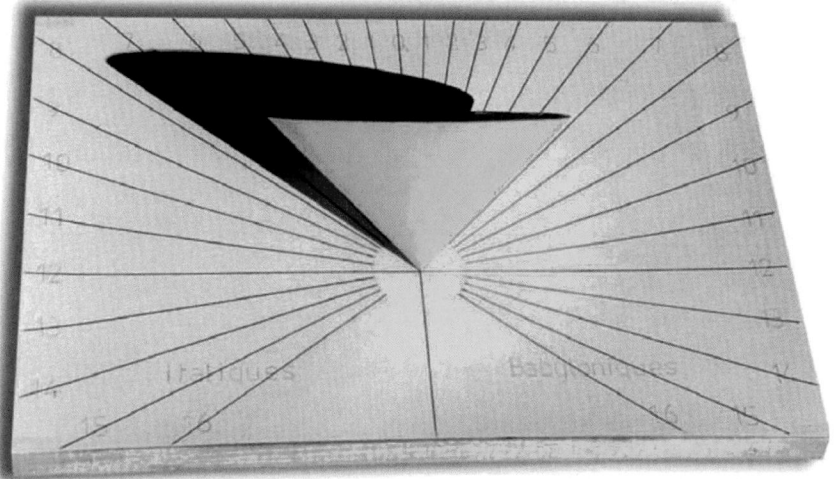

Figure 5.11 A dial showing Babylonian and Italic hours. The maker has shown here the amount of time in hours before sunset by modifying the Italic hours. (M. Vercasson)

intersection of the hour lines, and one of the generatrices of the cone lies on the 0 h line (the meridian).

Example

When the sundial is correctly aligned, what do we observe? When the Sun is shining, the shadow of the cone can tell us two things. The left-hand (western) edge of the shadow shows the Italic hours, and the right-hand (eastern) edge the Babylonian hours. Suppose for example that it is solar noon on 21 June, at latitude 49°. On the right-hand side of the dial we read 8 h (Babylonian): the Sun rose at about 4 h (true solar time) in the morning, 8 hours before solar noon. On the left-hand side, we read 16 h (Italic): the Sun set the previous day at 20 h (true solar time), so, at noon on the following day, 16 hours have passed since sunset.

This original sundial has great aesthetic appeal, but reading the time is quite difficult between 0 h and 4 h, since the relevant lines lie beneath the cone.

6 Polar sundials

The polar sundial is universal, and its easily realized diagram is completely independent of local latitude.

6.1 The principle of the polar sundial

The dial
The polar sundial is an inclined dial with parallel hour lines. As its name suggests, the table of the sundial is flat, and parallel to the polar axis. In other words, it is set at an angle to the meridian equal to the local latitude. As for the style, it is parallel to the table and in the plane of the meridian; the style can be the top edge of a rectangle mounted at right angles to the table, or a rod supported at both ends (as in Figures 6.1 and 6.2).

Hour lines
The hour lines are drawn parallel to the style. They are spaced according to the law of tangents (Figure 6.5): the further they are from the noon line, the further apart they will be, such that the 6 h and 18 h lines are theoretically at infinity. The pattern of lines is therefore limited to, at most, 7 h to 17 h. Morning hour lines are on the western side, and afternoon lines on the eastern side. The noon line is in the plane of the meridian. Since the diagram for the polar sundial is independent of local latitude, this kind of sundial will function equally well from the North Pole to the South Pole, using the same layout of lines. The only requirement is that the table be correctly inclined to the meridian so that the style points towards one of the Poles.

Marking out the polar sundial is easy. It can be very quickly accomplished using a pocket calculator. It is also possible to use the equatorial sundial as a projection system: while the table of the equatorial sundial is parallel to the Celestial Equator, the table of the polar sundial is parallel to the polar axis, such that the two tables are at right angles to each other (Figure 6.3).

The shadow of the style
Since the style is parallel to the dial, the shadow moves across the lines successively from west to east. If the style carries a marker (for example, a small ball), it can indicate certain dates: the shadow of the ball, as it moves across the surface of the dial, will describe hyperbolae. We could, for example, represent the two

Figure 6.1
A polar sundial
in the park of
the town hall in
Châteaubernard
(Charente).
(S.Grégori)

hyperbolae marking the solstices and the straight line marking the equinoxes (see
Figure 6.4, and the fact box on page 84).

Further information:

• Indicating the dates of the equinoxes and solstices on a polar sundial: see fact box on
  page 84.

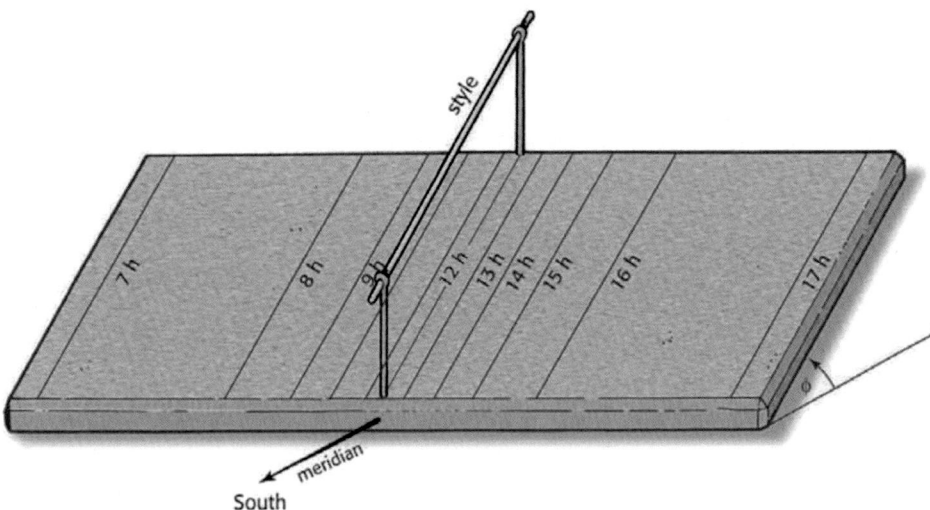

Figure 6.2 A polar dial. The table is plane and makes an angle φ with the meridian equal to the local latitude. The style is parallel to the dial: it points towards the north celestial pole. The noon line and the style are in the plane of the meridian.

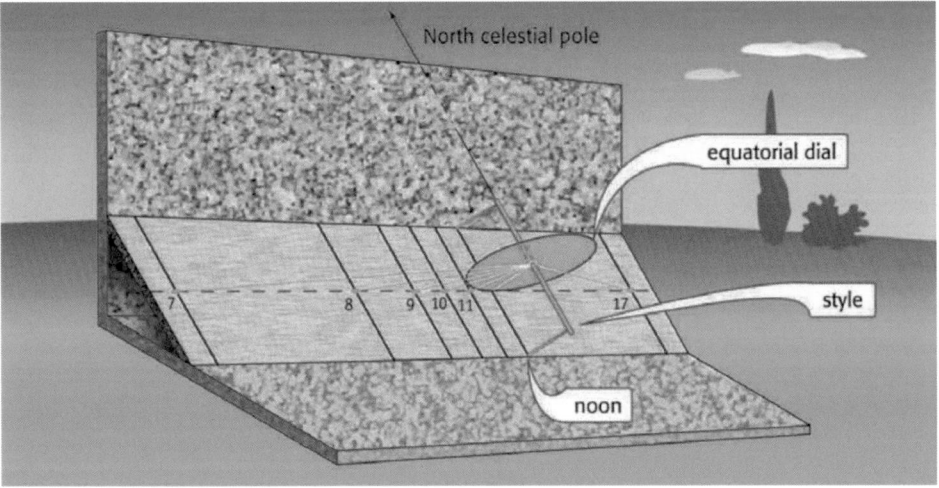

Figure 6.3 Hour lines on a polar dial. The hour lines of an equatorial dial have a regular 15° spacing. The hour lines on a polar dial are prolongations of those of a perpendicular equatorial dial whose centre coincides with the style of the polar dial.

Indicating the dates of the equinoxes and solstices on a polar sundial

As we have seen, a horizontal sundial for the Equator ($\varphi = 0°$) is in fact a polar sundial with parallel lines. To ensure that it is polar at other latitudes, we incline it such that its plane remains parallel to the Earth's axis. We shall use this property to draw the arcs for the solstitial hyperbolae and the straight line representing the equinoxes. We saw in the chapter on the horizontal sundial that the equation for the hyperbola, i.e. the curve described by the tip of the shadow of a gnomon, in the form $y = f(x)$, is written:

$$y = \frac{-a \sin \varphi \cos \varphi + \sin \delta \sqrt{x^2(\cos^2 \varphi - \sin^2 \delta) + a^2 \cos^2 \delta}}{[\sin^2 \delta - \cos^2 \varphi]}$$

where a is the length of the gnomon, φ the local latitude and δ the declination of the Sun. In the case of the polar sundial, we need only to make $\varphi = 0°$. Then, the formula becomes

$$y = -\tan \delta \sqrt{x^2 + a^2}$$

We place a bead (or a nut) in the middle of the polar style of the sundial: the shadow of the bead will describe a hyperbolic curve on the dial table. The origin of the coordinates will be the point below the bead on the noon line, with the x-axis towards the east and the y-axis towards the north. The hyperbola for the summer solstice is symmetrical with that for the winter solstice, to either side of the straight line representing the equinoxes.

 Example: if $a = 4$ cm, and $\delta = +23°.44$ and $x = 4$ cm, then $y = -2.45$ cm. If $\delta = 0°$, and $x = 4$ cm, then $y = 0$ cm. If $\delta = -23°.44$ and $x = 4$ cm, then $y = +2.45$ cm.

6.2 Marking out the sundial

Theory

On a polar sundial, the hour lines are marked out according to a simple trigonometrical relationship. Working from the noon line, they are spaced according to the law of tangents (Figure 6.5). The distance of an hour line from the noon line is equal to $a \tan H$, where a is the distance of the style from the table, and H is the hour angle of the Sun (1 h $= 15°$, 2 h $= 30°$, etc.). Let d be the distance between an hour line and noon, and let us assume $a = 1$. Table 6.1 gives the calculated distance for the various hours.

 We can confirm that d increases more and more as we move away from noon, in accordance with the law of tangents.

Table 6.1 The distances d of hour lines from the noon line for a style positioned one unit of length above the dial

Hour angle H	Distance d
15°	0.268
30°	0.577
45°	1
60°	1.732
75°	3.732
90°	infinity

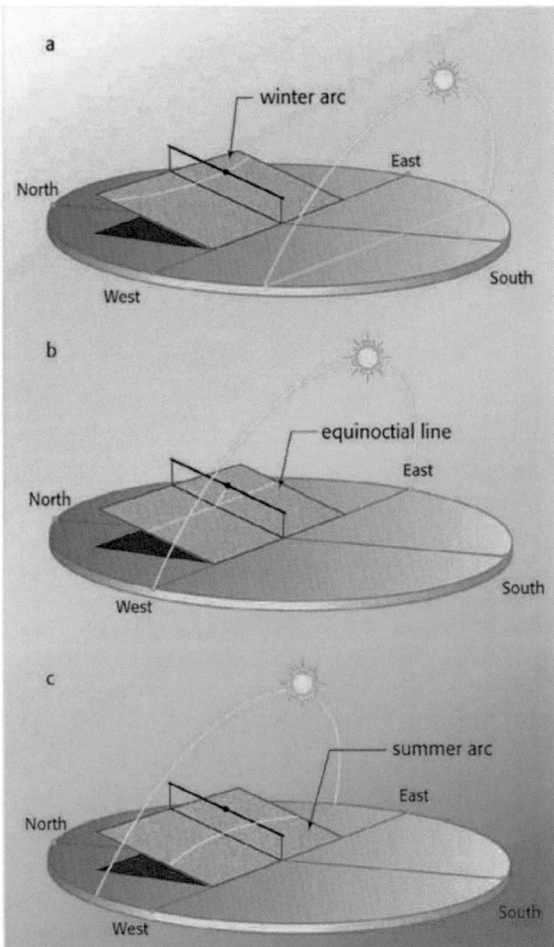

Figure 6.4 A polar dial at the winter solstice (a), the summer solstice (c), and at the equinoxes (b). Using a marker on the style, it is possible to use the polar dial as a calendar. The curves followed by the shadow of the marker in the course of a day are hyperbolae (like the summer and winter arcs), except at the equinoxes when the shadow follows a straight (equinoctial) line.

The morning lines (to the left of noon) are symmetrical to the afternoon lines (to the right). The 6 h and 18 h lines are at infinity. The spacing of the lines therefore is very dependent upon the distance a between the style and the dial table: if this distance is too great, only a few lines can be marked. Conversely, if the style is too close to the table, the hour lines will be too crowded.

We can calculate the height of the style *a priori* as a function of the size of the table and the number of hour lines which we wish to represent. If L is the distance between the noon line and the edge of the dial, the height a of the polar style above the table will be equal to $(L/\tan 75°)$ for the dial to indicate the time between 7 h and 17 h. If $L = 15$ cm, height a will be approximately 4 cm.

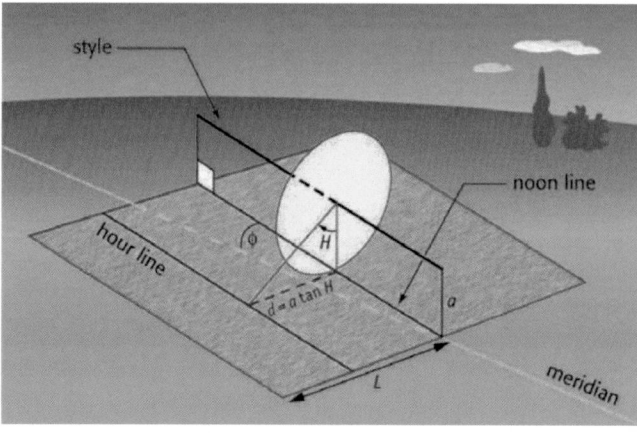

Figure 6.5 The law of tangents. The hour line corresponding to the hour angle H is at a distance of $d = a \tan H$ from the noon line.

Method

- **Materials required:** rectangular board (30 cm × 15 cm); wooden rectangle (4 cm × 15 cm), or 2 rods 4 cm long and a thick thread; protractor.

We take a rectangular board of dimensions 30 cm × 15 cm. Across the middle of the board, we mark, parallel to the shorter edge, a straight line representing the noon line (Figure 6.6). For the style, we can for example set up along the noon line, perpendicular to the plane of the dial table, a wooden rectangle or open rectan-

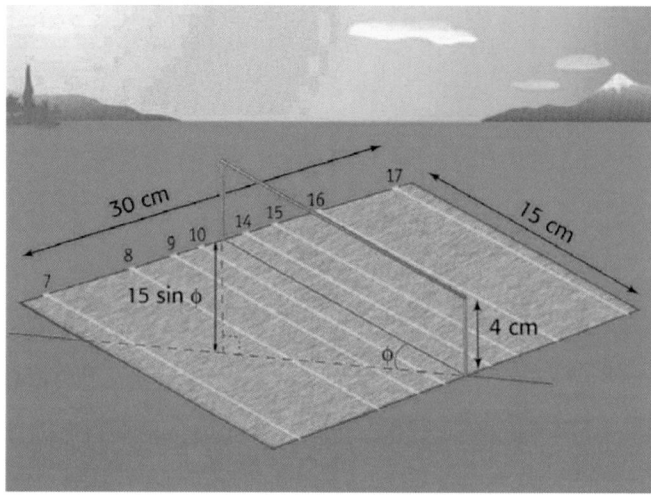

Figure 6.6 The construction of a polar sundial.

gular frame. In order to achieve the correct inclination of the dial table in accordance with the local latitude, we construct, for example, a wooden right-angled triangular wedge: the angle between the hypotenuse of the wedge and the ground will be equal to the local latitude φ. If the wedge is as long as the shorter edge of the dial (here 15 cm), these dimensions will be: opposite side $= 15 \sin \varphi$, adjacent side $= 15 \cos \varphi$.

Table 6.2 The distances d of hour lines from the noon line for a style positioned 4 cm above the dial

Hour line	d (cm)
13 h	1.1
14 h	2.3
15 h	4
16 h	6.9
17 h	14.9

Example

Let us imagine that our polar sundial is to be set up at latitude $\varphi = 48°$. The distances d of the hour lines from the noon line will be obtained using the relationship $d = a \tan H$. If we take a to be 4 cm, we obtain the distances indicated in Table 6.2.

In the case of the morning lines, we use the same distances, but to the left of the noon line. We therefore obtain a set of parallel lines across which the shadow of the style will move. If we construct a wooden wedge of hypotenuse 15 cm long, the dimensions of the triangle will be 11.1 cm for the opposite side, and 10 cm for the adjacent side.

When the dial has been completed, it remains only to place it with the noon line exactly upon the local meridian (see Section 2.5).

7 Vertical sundials

Vertical sundials are the type most often encountered, and are found on the walls of older houses, and on historic buildings, churches and monuments. They are often embellished with mottoes.

7.1 Vertical sundials

Unlike the other sundials already described, vertical dials exhibit an important peculiarity: normally, the walls on which they are displayed do not face due south. A sundial thus mounted is known as a declining dial (Figure 7.1): it may be a morning dial if the wall faces south-east or east, or an afternoon dial if facing south-west or west. Determining the orientation of a wall, and marking out a declining vertical dial and setting up its style, necessitate calculations which are outside the scope of this book.

Therefore, we shall concentrate only on those dials which face exactly towards one of the four cardinal points, as these are easy to make. We shall see that it is still possible to make a sundial for a wall, even if it does not face due south, with a little thought. It must be said that if we consider the wall to be south-facing, and this is not actually the case, our readings will be considerably adrift. For example, if we are 10° out in our estimate of the orientation, the dial may be in error by as much as 40 minutes! Also, if we make a sundial which is meant to face south and put it on a south-east-facing wall, the error may run into several hours!

Here, we suggest the construction of four dials, to face south, north, west and east. If we wish, we can mount them all on a single wooden cube, with a dial attached to each vertical face, and a horizontal sundial on the upper face. This kind of project is ideal to assist our understanding of the apparent motion of the Sun in the sky and the notion of the illumination of a plane.

7.2 Making a vertical direct south sundial

Theory

A vertical sundial which faces due south is called a direct south or meridional sundial, or a meridian. Its polar style makes an angle of $(90° - \varphi)$ with the vertical noon line. If we stand facing the dial, the shadow of the style moves from left to right during the course of the day, passing one by one across the hour lines, which

Figure 7.1 A declining vertical sundial at Mondovie (Italy). The dial table is vertical. The pattern of hour lines and the position of the style depend on the orientation of the dial. (S.Grégori)

all converge on the foot of the style. This kind of dial is always limited to the hours between 6 h and 18 h (Figure 7.2).

Method

- **Materials required:** wooden square (35 cm × 35 cm); wooden rectangle; protractor.

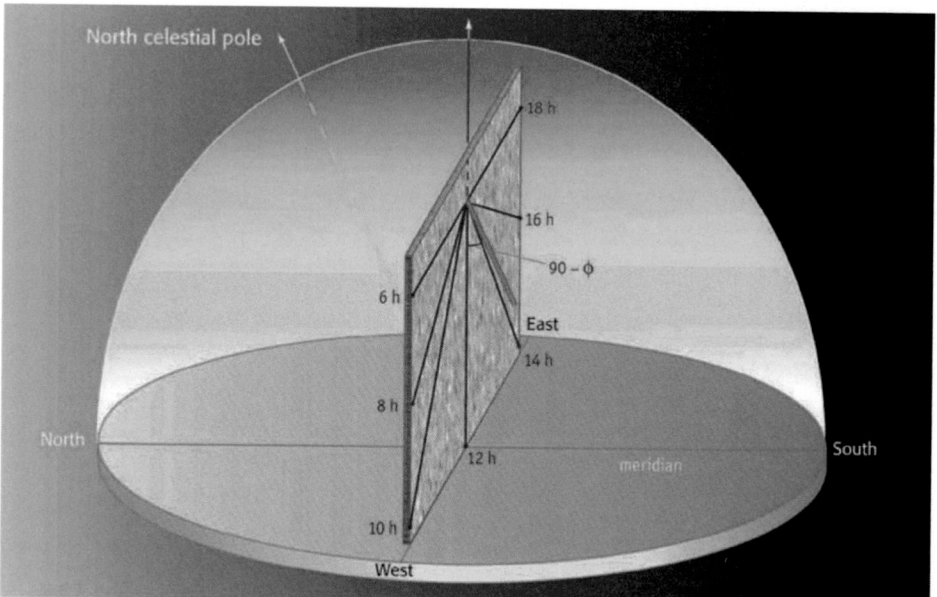

Figure 7.2   A meridional dial. The style is polar. On the vertical table, facing due south, are the hour lines from 6 h to 18 h.

Within the wooden square, we mark out another square of 30 cm × 30 cm, leaving enough room around the outside to write the hour numbers. We draw a vertical line from the half-way point of one of the sides of the inner square; this line becomes our noon line. The intersection of the noon line with the upper edge of the inner square represents the point at which we will insert our style (Figure 7.3).

From this point, we then draw angles H' between the hour lines and the noon line, as on the horizontal sundial. H' can be calculated using the relationship:

$$\tan H' = \cos \varphi \tan H$$

where H is the hour angle of the Sun ($H = 0°$ at noon, $H = 15°$ at 13 h, $H = 30°$ at 14 h, ..., $H = -15°$ at 11 h), and φ is the local latitude. Since a vertical direct south dial is limited to the hours between 6 h and 18 h, we make $H' = \pm 90°$ if $H = \pm 90°$, as in the case of the horizontal sundial. Consequently, the 6 h and 18 h lines are perpendicular to the vertical noon line (Figure 7.4). Table 7.1 shows calculated values for $\varphi = 48°$. Morning angles are symmetrical to their afternoon counterparts. The style points towards the North Celestial Pole, and the angle between the style and the noon line is equal to $(90° - \varphi)$, here 42°. To make the style, we can cut a triangle from a wooden rectangle: the hypotenuse will represent the style itself.

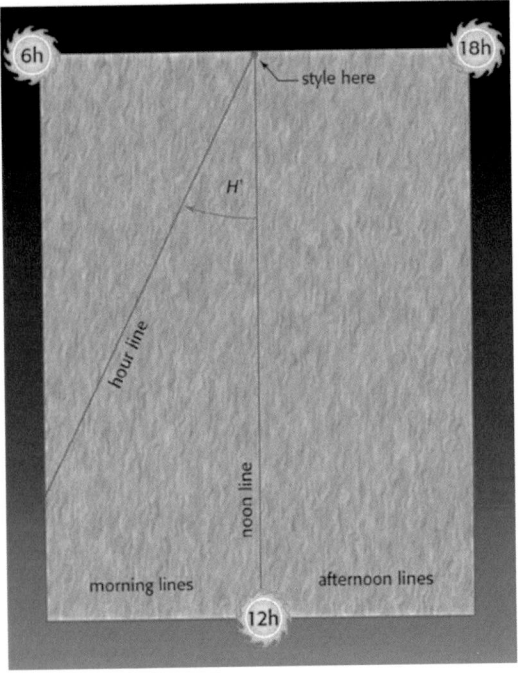

Figure 7.3 Drawing the hour lines on a vertical meridional sundial. The morning lines are on the western side of the dial and the afternoon lines are on the eastern side.

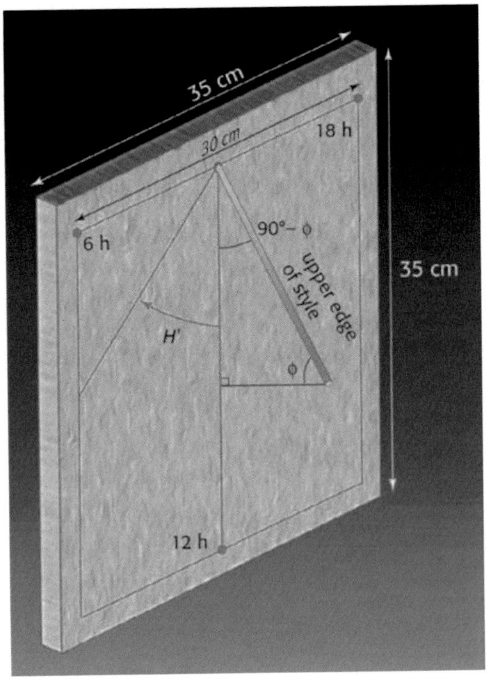

Figure 7.4 The construction of a meridional sundial.

It should be noted that, if the style is at all thick (see above, in the section on the horizontal sundial), the afternoon hour lines converge at the right-hand side of the foot of the style, and the morning lines to the left-hand side.

Practical example

Let us imagine a house with one of its walls facing south-west or south-east. With a little thought we can mount a sundial on one of these walls. We construct a direct south dial, but instead of fixing it directly onto the wall, we angle it outwards so that it faces due south (Figure 7.5). The dial must therefore pivot around its eastern edge if the wall faces south-west, or around its western edge if the wall faces south-east. In order to set up the dial facing south, we work out the moment when the Sun crosses the meridian (see Chapter 1): then, the shadow of the style must indicate solar noon. We need only await that moment, swing the dial around so that it indicates noon, and then fix it in place.

Calendar

In the course of a day, the shadow of the tip of the style will follow a hyperbolic path, except at the equinoxes, when it moves in a straight line (the equinoctial line, see Figure 7.6). It is therefore possible to draw the hyperbolae from known equations for different noteworthy dates (see fact box on page 94), and use the meridional dial as a calendar.

Further information:

- For how long will a direct south sundial be illuminated? see Appendix F, page 162. Formula for the direct south sundial based on the horizontal sundial: see Appendix E, page 156.

Table 7.1 Hour line angles H' from the noon line for a meridional dial at latitude $\varphi = 48°$.

Hour angle H	Hour line angle H from noon line
0°	0°
15°	10°.16
30°	21°.12
45°	33°.79
60°	49°.21
75°	68°.18
90°	90°

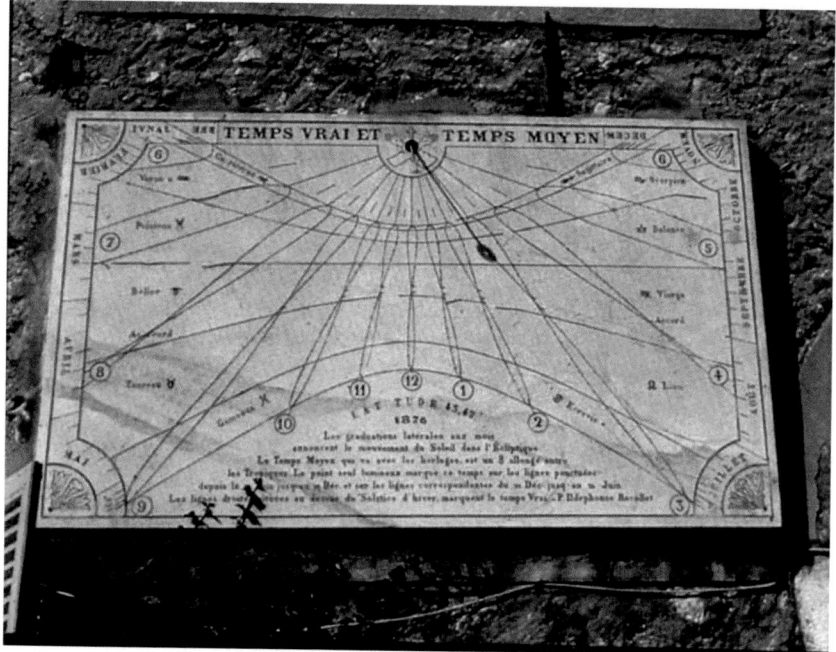

Figure 7.5 A vertical dial on the convent at Cimiez (Nice). On each hour line, we see a figure-of-eight curve, allowing mean time as well as solar time to be read. (SAF/R.Sagot)

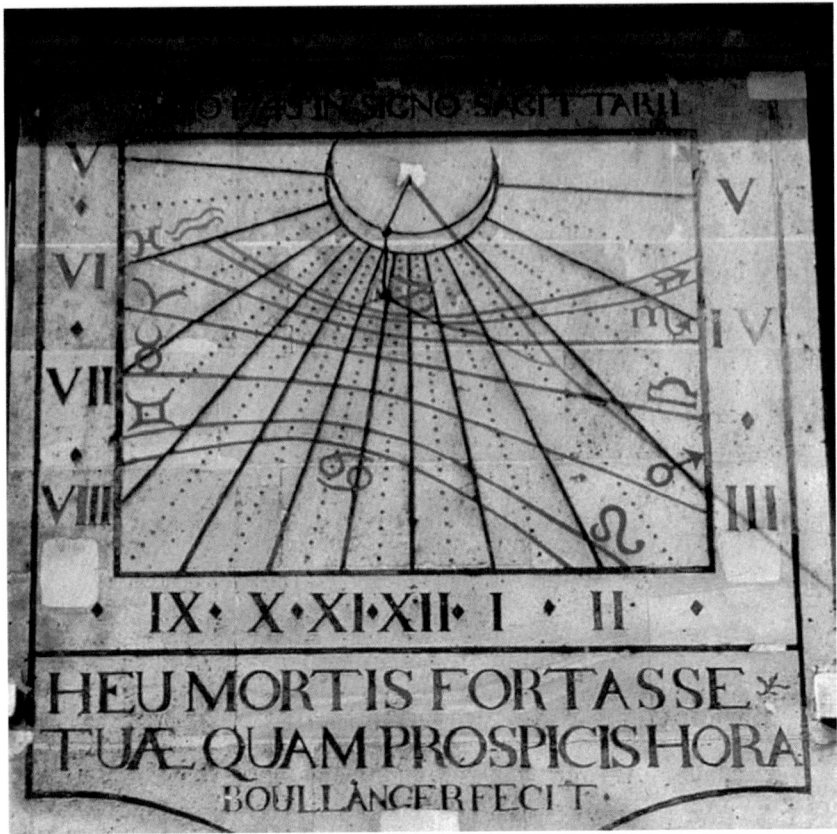

Figure 7.6 Vertical (slightly) declining sundial at the Hôpital Laënnec in Paris. The zodiacal curves also appear. (SAF/D. Savoie)

Drawing the hyperbolae for a direct south (meridional) sundial

Before proceeding, it must be pointed out that the drawing for the face of a direct south sundial at latitude φ is the same as that for a horizontal sundial at latitude $(\varphi - 90°)$. We saw in the chapter on the horizontal sundial (Chapter 5) that the equation for the hyperbola, i.e. the curve described by the tip of the shadow of a gnomon, in the form $y = f(x)$, is written:

$$y = \frac{-a \sin \varphi \cos \varphi + \sin \delta \sqrt{x^2 (\cos^2 \varphi - \sin^2 \delta) + a^2 \cos^2 \delta}}{[\sin^2 \delta - \cos^2 \varphi]}$$

where a is the length of the gnomon, φ the local latitude and δ the declination of the Sun. We can use this formula to draw the hyperbolae on a direct south sundial, with two changes. The first involves the length of the gnomon a, which on a direct south sundial is equal to ($U \cos \varphi$), where U is the length of the polar style. The origin of the series of points is therefore that point beneath the foot of the polar style and at distance ($U \sin \varphi$) from it, with the x-axis towards the east and the y-axis towards the zenith. Finally, we need to replace φ in the formula by ($\varphi - 90°$).

   Example: our direct south sundial is set up at $\varphi = 48°$. Its polar style is 10 cm long. The length of the gnomon is therefore 6.69 cm and it is situated at 7.43 cm on the noon line below the foot of the polar style. In the formula, we enter the value $\varphi = -42°$. If $\delta = +23°.44$ and $x = 9.47$ cm, then $y = -17.1$ cm. If $\delta = -23°.44$ and $x = 3.58$ cm, then $y = -1.84$ cm.

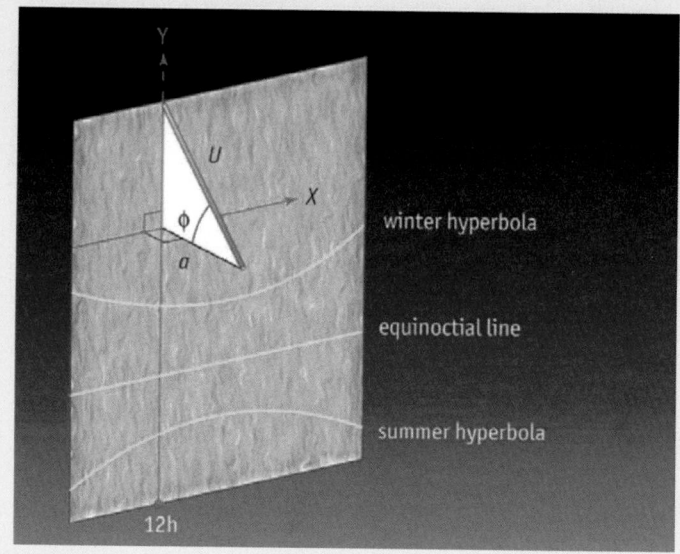

7.3 Making a vertical direct north sundial

Theory

It is a common belief that, at the latitude where we live, the northern side of a house is always in shadow. However, we can easily see for ourselves in spring and in summer, between the two equinoxes, that the Sun's light falls upon this side for a short time in the early morning and late afternoon. We can see from this fact that a sundial which faces north will work for nearly six months every year. At the June solstice, it will work for the longest period during one day. Obviously, a north-facing wall will not be illuminated at midday (in Europe)! Making a direct north (or septentrional) dial involves drawing a dial on the opposite face of a direct south dial. In this case, the style will point upwards, since the North Celestial Pole makes an angle with the horizon equal to the latitude (Figure 7.7).

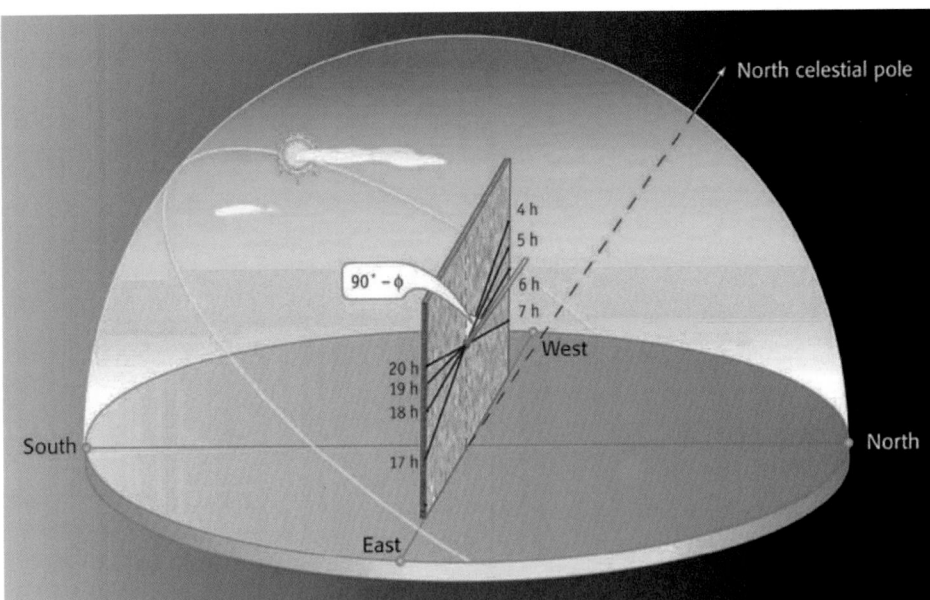

Figure 7.7 A septentrional dial. The style points towards the North Celestial Pole. The dial is illuminated by the Sun between the spring equinox and the autumn equinox. It carries eight hour lines, from 4 h to 7 h and from 17 h to 20 h.

If we stand facing a direct north dial, the shadow of the style will move in a clockwise direction around the table. At our latitude, we can draw eight hour lines: four for the morning, from 4 h to 7 h, on the right-hand side of the dial, and four on the left-hand side, from 17 h to 20 h. All the lines converge on the foot of the style (Figure 7.8).

Method

- **Materials required:** wooden square (35 cm × 35 cm).

As in the case of the direct south dial, we draw an inner square of 30 cm × 30 cm on the wooden board, leaving enough room around the outside to write the hour numbers (Figure 7.9). We draw a horizontal line from the half-way point of one of the sides of the inner square. We draw a vertical line through the midpoint of this horizontal line, and the intersection of these two lines is where we place the polar style. The style is set at an angle with the vertical board of $(90° - \varphi)$.

The number of hour lines available is fairly limited. At our latitude, we can mark lines for 4 h, 5 h, 6 h and 7 h for the morning, and 17 h, 18 h, 19 h and 20 h for the afternoon. Note that the 19 h line is a prolongation of the 7 h line, the 5 h line is a prolongation of the 17 h line, and so forth (Figure 7.10).

Figure 7.8　A septentrional sundial. This dial is drawn on the north face of a cube. (SAF/ R.Sagot)

The hour lines are drawn with reference to the vertical line by using formula (1). Take care, since the sign of H' must be the same as that of H, which means that we must sometimes add or subtract $180°$.

Example

Let us give a complete calculation for $\varphi = 48°$. Table 7.2 lists the hour angles for morning and evening.

Further information:

- On what dates is a vertical direct north sundial sunlit, and for how long? See Appendix F, page 162.

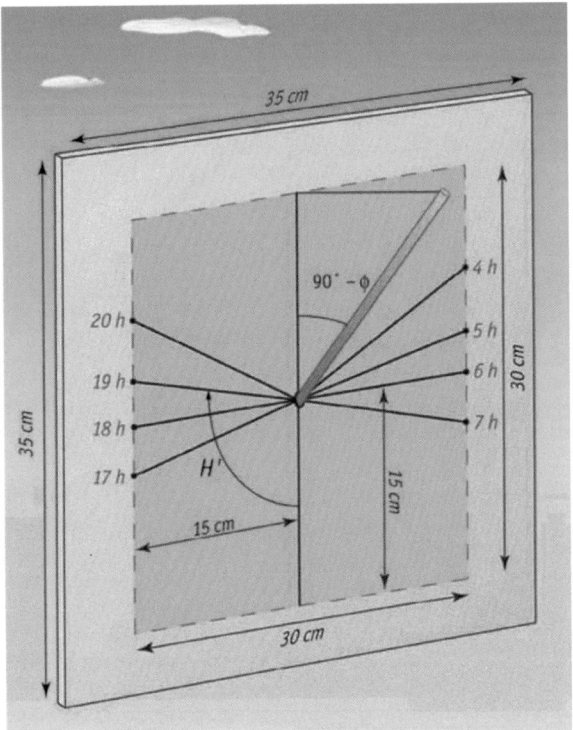

Figure 7.9 The construction of a septentrional sundial.

7.4 Making vertical direct west and direct east sundials

Theory
A sundial facing due west or due east has something in common with a polar sundial: its style is parallel to the surface of the dial, and points at the celestial pole, while its hour lines are parallel to the style (Figure 7.11).

A direct east dial will indicate the time from sunrise until the Sun passes beyond the plane of the wall, and a direct west dial from the time of the appearance of the Sun from beyond the plane of the wall until sunset. Neither of these dials will be able to indicate noon.

The direct west sundial: method
- **Materials required:** wooden board (35 cm × 35 cm); some metal rods.

A direct west sundial must be set up exactly facing the +90° (due west) point on the horizon. It is very easy to make. Within the 35-cm wooden square, we mark out another square of 30 cm × 30 cm, leaving enough room around the outside to write

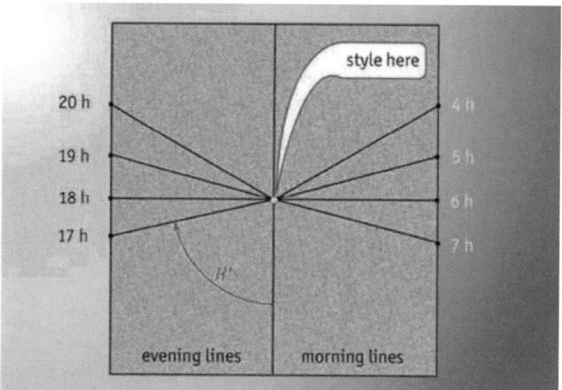

Figure 7.10   Drawing the hour lines on a septentrional sundial. The evening lines are on the eastern side of the dial and the morning lines are on the western side.

Table 7.2   Hour line angles H' from the noon line for a septentrional dial at latitude $\varphi = 48°$

Solar time	Hour angle H	Hour line angle H' from noon line
4 h	−120°	−130°.79
5 h	−105°	−111°.82
6 h	−90°	−90°
7 h	−75°	−68°.18
17 h	75°	68°.18
18 h	90°	90°
19 h	105°	111°.82
20 h	120°	130°.79

the hour numbers (Figure 7.13). We draw a horizontal line 8 cm from the upper edge of the inner square. We mark a point P on this line, 8 cm from the right-hand edge of this square. We draw a straight line, pointing towards the North Celestial Pole, through point P, at an angle with the horizontal line equal to the local latitude. The style will be located above this line and parallel to it; it could be mounted as shown in Figure 7.12, supported at both ends by rods. Its distance a from the dial table will determine the spacing of the hour lines. If we chose, for example, to mount it 8 cm from the table, we would achieve a maximum number of hour lines for the dimensions given.

We now draw, at right angles to the line pointing towards the pole, a straight line passing through P. The hour lines will be at right angles to this line, each at a distance $(a/\tan H)$ from point P. The 18 h line will lie beneath and parallel to the style. During the afternoon, the shadow of the style will pass successively across the hour lines, remaining parallel to them.

Table 7.3 gives, for latitude 48°, the distances of the various hour lines from point P, if $a = 8$ cm. The 19 h and 20 h lines will be above point P on the vertical dial, and the others below it.

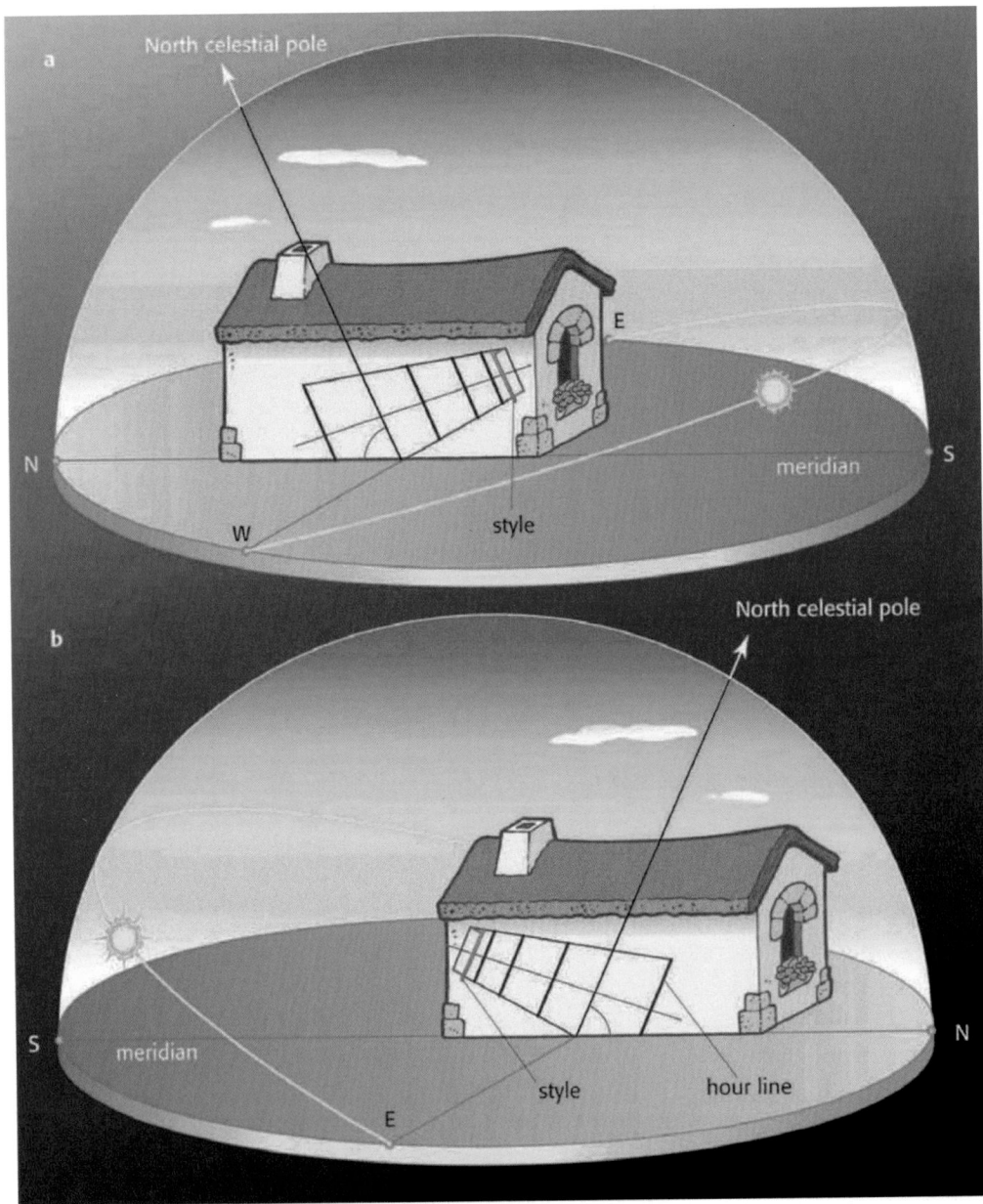

Figure 7.11 Direct west (a) and direct east (b) sundials. The table is orientated towards the west (a) or the east (b). The styles, which are parallel to the dial, point towards the North Celestial Pole.

Figure 7.12 Vertical sundial, facing approximately east, at Peveragno (Italy). (S.Grégori)

The direct east sundial: method

- **Materials required:** wooden board (35 cm × 35 cm); some metal rods.

A direct east sundial is a reversed version of the sundial we have just described. The dial is set up exactly facing the −90° (due east) point on the horizon. It will show the time from sunrise to noon (though the noon line is theoretically at infinity). To mark it out, we follow the instructions for the direct west dial, with a few modifications. On the 30 cm × 30 cm square, point P is situated 8 cm from the upper edge and 8 cm from the left-hand edge (Figure 7.14). The straight line passing through P makes an angle with the horizontal equal to the local latitude. The hour lines will be drawn parallel to a straight line at right angles to the line pointing towards the pole, again using the relationship $(a/\tan H)$. The 6 h line will lie beneath and parallel to the style. The 4 h and 5 h lines will be above P on the vertical dial. Table 7.4 gives, for latitude 48°, the distances of the various hour lines from point P, if $a = 8$ cm.

Conclusion

The four sundials described above (direct north, south, west and east) can indicate solar time only if they are correctly oriented. We could envisage marking them out, for example, all on the faces of a cube (Figure 7.15): four vertical sundials, and perhaps a horizontal sundial on the upper face. Since the four dials will all move in

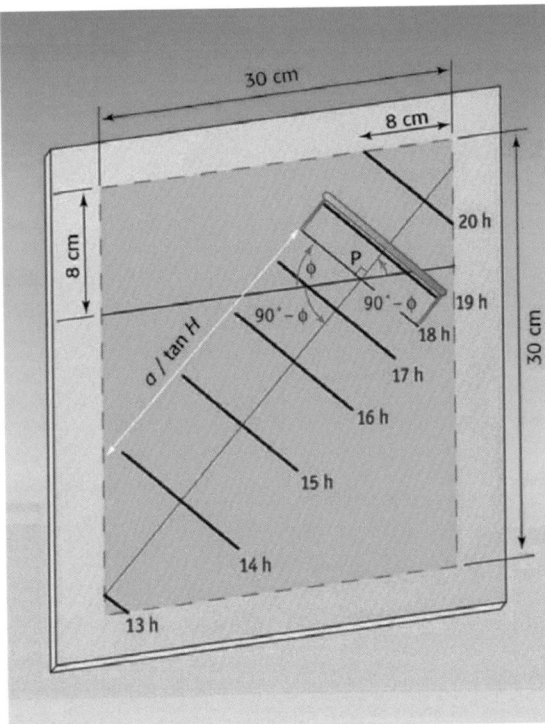

Figure 7.13   Making a direct west dial. The style is supported above the table and parallel to it at distance a. It makes an angle φ with the horizontal equal to the local latitude. The hour line corresponding to the hour angle H is at a distance of a/tan H from the style.

unison, we have only to turn the cube until three of the dials show the same time in order to orient it exactly with the local meridian!

7.5   Making a reflection sundial

Theory
There are very few reflection sundials, one of the best known being that of the Lycée Stendhal in Grenoble, dating from 1673. The principle of the reflection sundial is as follows: a small mirror placed near a window throws a spot of light into a room decorated with a pattern of lines and curves, and this shows the time. Instead of the shadow of a style being used to indicate the hour and the date, the spot of light performs these functions as it travels through a shaded area. So here we have an unusual dial based upon the law of reflection.

Empirical method
This kind of sundial can be set up with relative ease, and is a useful exercise in comprehension. First, we have to determine the best place to site the mirror. This

has to be a matter of trial and error, since the room into which the light is to be projected will probably not face due south; it may not even be possible for the dial to show solar noon. We can also experiment with the mirror's inclination and orientation to achieve the best position.

Once the mirror is in place, it projects into the room a spot of light which will move along the walls and the ceiling. We proceed empirically to mark out the dial within the room, referring to the position of the spot of light at various times of the year. Obviously, we consult a clock in order to work out solar time every day, or our

Table 7.3 Positions of hour lines on a direct west dial for $\varphi = 48°$ and $a = 8$ cm

Solar time	Hour angle H	Distance of hour line from point P
13 h	15°	29.9 cm
14 h	30°	13.9 cm
15 h	45°	8.0 cm
16 h	60°	4.6 cm
17 h	75°	2.1 cm
18 h	90°	0.0 cm
19 h	105°	2.1 cm
20 h	120°	4.6 cm

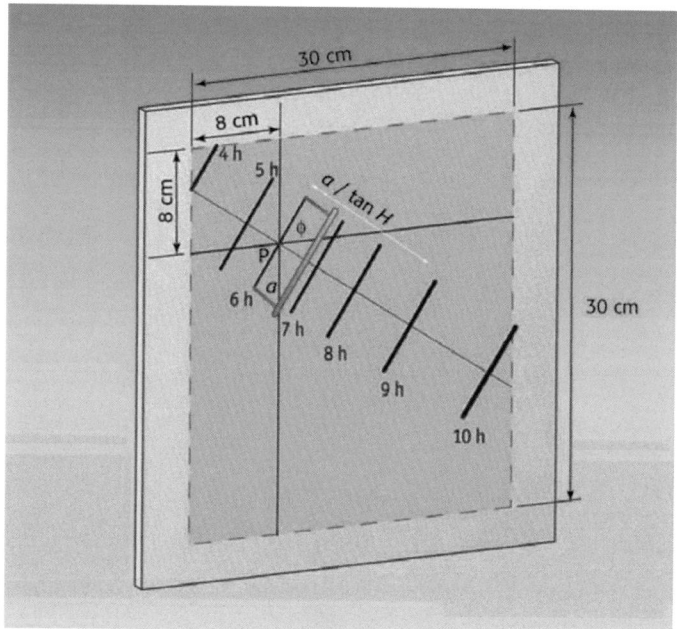

Figure 7.14
Making a direct east sundial.

Table 7.4 Positions of hour lines on a direct east dial for $\varphi = 48°$ and $a = 8$ cm

Solar time	Hour angle H	Distance of hour line from point P
4 h	−120°	4.6 cm
5 h	−105°	2.1 cm
6 h	−90°	0.0 cm
7 h	−75°	2.1 cm
8 h	−60°	4.6 cm
9 h	−45°	8.0 cm
10 h	−30°	13.9 cm
11 h	−15°	29.9 cm

efforts will be in vain. To this end, we need to know the local longitude and the equation of time (see Chapter 5, on the horizontal sundial, where the method is explained).

As the days go by, we build up a network of points which we need only to join up. For example, all the points corresponding to 11 h solar time will form the 11 h line. If we have the opportunity to work on our dial at the equinoxes and solstices, it is also possible to join all the points for these dates to mark out the curves representing the seasons.

Calculations for a vertical direct south reflection sundial

In the configuration shown in Figure 7.16, the hour lines all converge towards the meridian. The dial shows only the hours between 6 h and 18 h.

Down the middle of a rectangular wooden board, we draw a straight line (which will be the noon line). At right angles to this line we fix a horizontal projecting ledge to which we attach a small mirror M, about the size of a wrist-watch. When the dial is oriented due south, the horizontal mirror M casts the image of the Sun onto the vertical face. By virtue of the law of reflection, the image of the Sun on the upper part of the vertical surface lies symmetrical to the position where it would be if the light passed through the mirror and fell upon the lower part, below the horizontal ledge.

The distance between the mirror and the vertical surface is important: if the mirror M is too far away, the spot of light can 'fall' outside the dial. It is therefore advisable to experiment a little before settling for the final configuration. For local latitude φ, angle H' between an hour line and the vertical noon line is calculated using

$$\tan H' = \cos \varphi \tan H$$

where H is the hour angle of the Sun: $H = 0°$ for noon, $H = 15°$ for 13 h, $H = 30°$ for 14 h, ..., $H = -15°$ for 11 h. The 13 h line, for example, lies to the right of noon, and the 11 h line to the left.

Unfortunately, this kind of dial has one drawback: the spread of the hour lines is limited, because the spot of light rapidly moves off the dial.

Figure 7.15
There are five sundials on this cube, four vertical and one horizontal. Three of the faces are indicating 9 h. These dials were created by the astronomer Leverrier (Agon, Manche). (SAF/ R.Sagot)

7.6 Making a sundial on a plane surface without calculation

Theory

We may wish to make a sundial for a wall which does not face due south, or on some inclined surface of whatever orientation. Often, we will be frustrated by the very complex calculations involved in such a project. We can however overcome this difficulty in a very simple way.

Mass-produced sundials: from rough guide to 'rip-off'!

For more than a decade now we have seen displays of sundials in shops and garden centers. To the curious and to holidaymakers, these dials can seem very attractive. They are usually made of stone or ceramic, and they often sport some motto, enhancing their 'retro' and authentic air. Such mass-produced, factory-made dials come in various forms, horizontal, vertical and even portable. Some dials are sold with a little sheet of explanatory notes.

Sadly, these sundials, set up on the walls and in the gardens of those who buy them, have one major defect: almost without exception, they don't tell the time! Leaving aside the fact that their owners are unaware that they must take into account the equation of time, local longitude and the difference from Universal Time, in order to convert solar time into clock time . . .

The purists among them may well know that, for example at the beginning of November, a sundial set up in Montpellier needs a correction of 28 minutes to show clock time. In fact, owners of sundials are often convinced that their instruments show the right time—a 28-minute error is negligible in such a venerable instrument! They sometimes move the style around to line up the shadow with the correct time, an operation which may have to be repeated many times . . .

Such 'factory' wall-mounted sundials—at least those which are correct—are generally adjusted for latitude 45° and a direct south orientation. All well and good for the owners, who believe that they have a 'south-facing' wall! In reality, a wall facing due south is a very rare thing, and a discrepancy of just one degree in the orientation can mean, at certain times of the year, an error on the dial of as much as four minutes; a discrepancy of one degree of latitude can introduce an error of up to two minutes.

Now, the owners of 'factory' sundials will soon realize that their instruments do not work properly, and the dials will become mere ornaments on the wall, adding to the 'olde-worlde' look of their houses. So we find horizontal sundials mounted on walls, direct south dials on east-facing walls, etc.

There are some who do try to understand what is wrong with their sundials: but they won't get much help from their sheet of explanatory notes! Not only do these notes often contain historical falsehoods, but they show a worrying lack of knowledge of gnomonics on the part of their authors.

Worse still, many mass-produced sundials are simple 'rip-offs': the angles of their styles are all wrong and the spacing between the hour lines owes more to sheer fantasy than to the laws of gnomonics. In other words, many of these dials are little 'tourist traps'.

The upshot of all this is that 'factory' dials give sundials in general a bad name, causing people to think that they are merely for decoration or that they will never work.

We might argue from an opposite standpoint, saying that the wide availability of sundials is bringing them back into fashion. There is some truth in this, but the ubiquitous displays of sundials, between the bird-tables and the wind chimes, only trivialize them and relegate them to the level of just another consumer item. Of course, everybody has the right to set up a sundial, but those who acquire them should buy in full knowledge of the facts: and that is where the difficulty arises.

Mass-produced sundial (Var). This sundial may be decorative, but is completely misleading as installed. (S.Grégori)

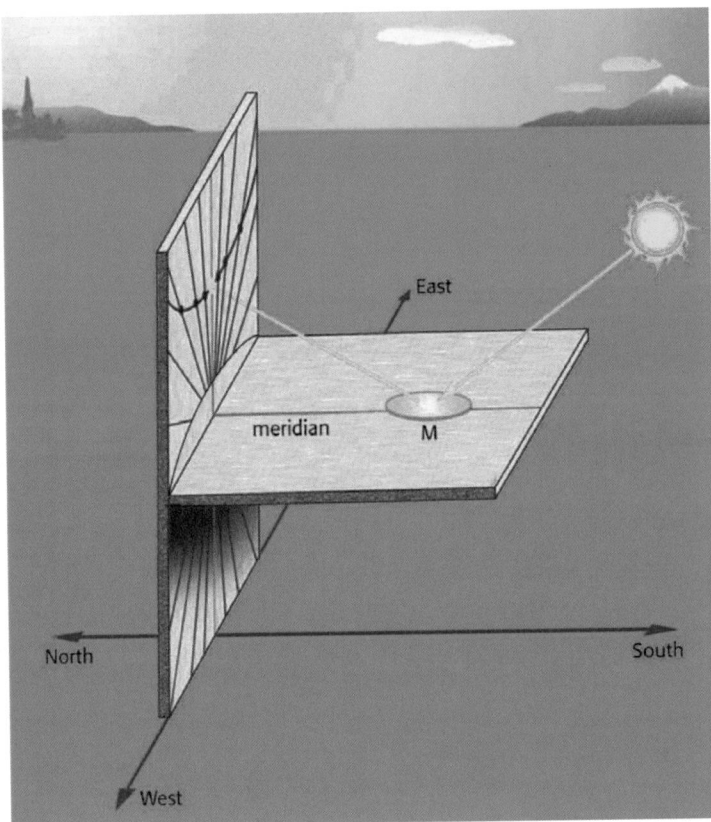

Figure 7.16
Direct south
reflection sundial.
The dial table
faces due south,
and the mirror M
is horizontal and
on the meridian.

Let us imagine that we can set the polar style (i.e. the rod which throws the shadow) directly into a surface of whatever orientation or inclination. To draw the hour lines on the as yet unmarked surface, we need only to use another sundial as a reference (when our correctly set-up reference dial shows a certain hour, we transfer that hour to the shadow of the polar style on the new surface). Alternatively, we can use the clock method (see section 5.6).

As for the direction in which the polar style points, we need to make it parallel to the Earth's axis of rotation; and to accomplish this, we shall draw our inspiration from a method dating back to the seventeenth century.

Method

• **Materials required:** stiff card; protractor; threaded rod 4 cm long set into a 1-cm thick wooden board (the rod is perpendicular to it).

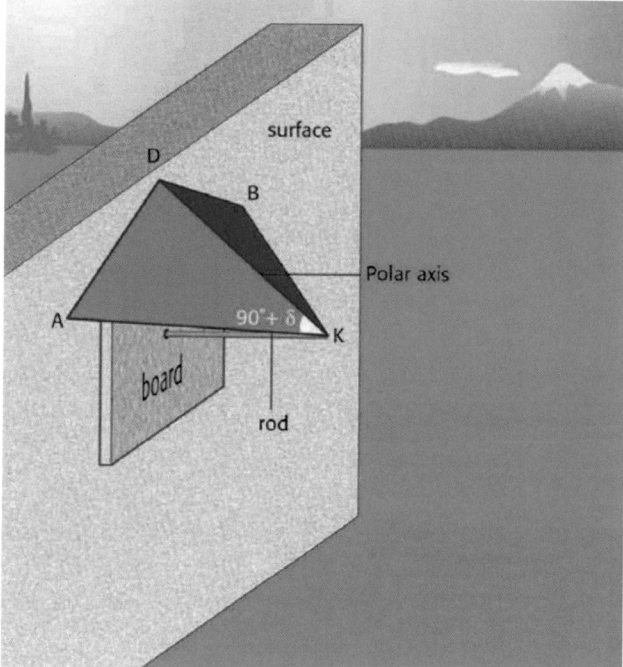

Figure 7.17 Determining the polar axis for any surface. The construction of the tetrahedron KDAB is carried out using the measurements KA and KB and the declination angle δ of the day in question. The polar axis is represented by the ridge KD.

To the surface on which the sundial is to be set up, we fix, temporarily, the threaded rod PK, set in its board. The rod will therefore be at right angles to the surface, and K will be the tip of the rod (Figure 7.17). We mark two points A and B where the tip of its shadow lies, one in the morning and one in the afternoon. We measure the distances KA and KB. From the table of solar declinations (see Appendix C, page 148), we determine the Sun's declination δ for the day in question. From the stiff card, we make a tetrahedron KDAB: we draw the triangles DKA and DKB, the lengths of whose sides KD (length arbitrary), KA and KB are known, as are the angles DKA or DKB:

$$\widehat{DKA} = \widehat{DKB} = 90° + \delta$$

We fold along KD and seat the folded card above the rod such that points K, A and B on the card coincide with the established points K, A and B. Now, the line KD is parallel to the Earth's axis. Now, we merely place a rod along the fold KD to obtain the polar style, which will be inserted into the surface of the dial.

In practice, there are several ways to increase the accuracy of this geometrical method. If astronomical ephemerides are not available, but only the table of declinations in Appendix C (see page 148), it is preferable to carry out this project around the time of the solstice, in a period when the declination of the

Sun changes very little. It is for this reason that angle DKA is considered here to be equal to angle DKB. In reality, the declination of the Sun varies between the two measurements, and it is more accurate to write:

$$\widehat{DKA} = 90° + \delta \quad \text{and} \quad \widehat{DKB} = 90° + \delta'$$

where δ is the value for the declination at the time of the first marking and δ' the declination at the second marking. There is nothing to prevent us from marking point A on one day and point B on another.

In theory, this method is independent of the orientation of the dial, its inclination, and the local latitude and longitude. All that we need to know is the Sun's declination. Once the style has been inserted, we must draw the hour lines, since a polar style by itself tells us nothing. As has been stated, we can use either a reference sundial or a clock: the latter method assumes that we know the local longitude and the equation of time.

8 Horizontal analemmatic sundials

The analemmatic sundial differs from those already described, essentially because the system of projection used in creating it is not a gnomonic one.

8.1 The analemmatic sundial

Projection systems

Just as there are different methods of cartographic representation, there are various systems involving projection which are used for tracing out different types of sundial. The gnomonic projection (Figure 8.1) takes as its starting point the centre of the Earth, and the circles on the Celestial Sphere are projected onto a plane tangent to the Earth's surface (Figure 2.10a, page 38). In this way we obtain straight lines (which explains why lines on traditional sundials are straight) for the great circles, and conics (hyperbolae at our latitude) for the lesser circles, such as the declination circles and the tropics. We can, of course, use any viewpoint for projection onto the plane tangent to the Earth. In the case of the analemmatic

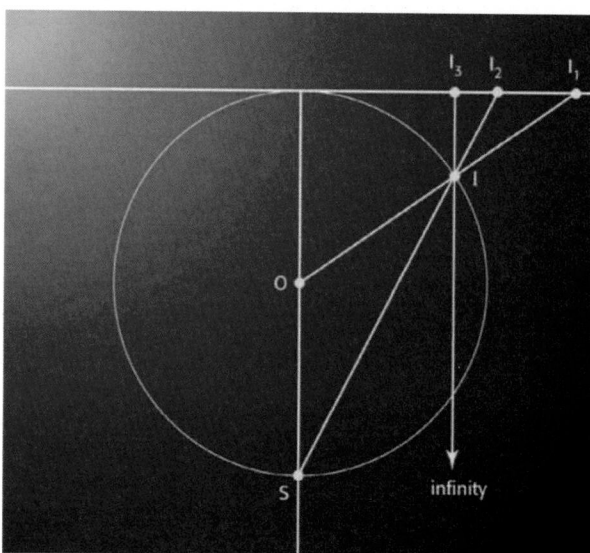

Figure 8.1 Gnomonic, stereographic and orthographic projections. For a projection on a plane surface, we may choose different points of view. Point I on the sphere can be projected onto the tangent plane from O (gnomonic projection to I_1), from S (stereographic projection to I_2), or from infinity (orthographic projection to I_3).

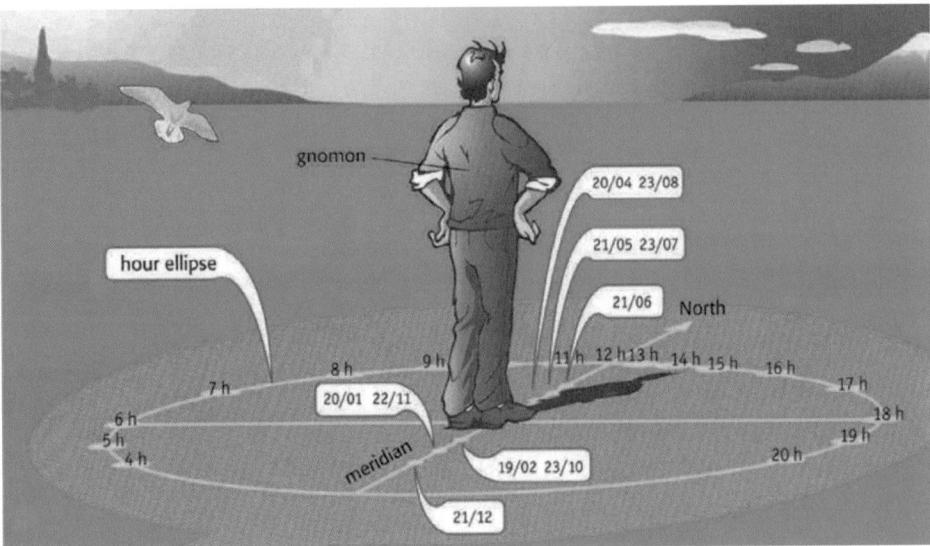

Figure 8.2 Horizontal analemmatic sundial. The hour points lie on an ellipse. The gnomon moves along the meridian as a function of the date and casts its shadow on the hour ellipse.

sundial, this viewpoint is at infinity: this type of projection is known as ortho-graphic. *Analemma* signifies an orthographic projection of the circles of the Celestial Sphere. On a plane tangent to the Earth, these are ellipses. The disadvantage of this projection is that we have to draw a different ellipse on the dial for every single day, which is hardly practicable. In the 17th century, a Frenchman called Vaulezard had the idea of constructing a single ellipse, and using a movable gnomon.

The theory of the analemmatic sundial
The classic analemmatic sundial is a horizontal dial. Instead of hour lines, it has hour points which lie on an ellipse known as the hour ellipse. This kind of sundial works from sunrise to sunset, i.e. for a maximum period of between 4 h and 20 h. The style is a movable vertical gnomon, which changes its position as a function of the date as a scale aligned with the local meridian. At a given moment, the shadow of the gnomon intersects the ellipse and solar time is indicated by the direction of the shadow (Figure 8.2).

8.2 Marking out the analemmatic sundial

The horizontal analemmatic sundial is particularly appropriate for a school playground: not only is it easy to mark out, but also a person can be used as a style! It can also be made on a smaller scale and integrated with a traditional horizontal sundial to obtain a veritable solar compass which indicates geographical north.

Drawing the ellipse

First, we have to determine the local meridian, on an absolutely horizontal surface, following the methods described in Chapter 2. Then we select a point of origin O on this meridian. Through O and at right angles to the meridian is the east-west axis. On the east-west axis we mark two points A and B, each 2 meters from O. The distance OA, like OB, is known as the semi-major axis of the ellipse. To the north of O we mark point C, the distance OC being equal to OA sin φ. This distance is known as the semi-minor axis of the ellipse.

Example 1: if φ = 46° and OA = 2 m, then OC = 1.44 m
To draw the ellipse, we may use the so-called gardeners' method, which is based on a geometrical property of the ellipse (Figure 8.3): an ellipse is the geometrical location of points, the sum of whose distances from two fixed points known as foci, is constant. Let us call the two foci F_1 and F_2. They lie on the major axis AB of the ellipse: the distance between O and one of the foci is equal to $\sqrt{(OA^2 - OC^2)}$, or OA cos φ. To each focus, we attach one end of a strong cord of length 2 OA. Then, with an appropriate marker, and keeping the cord taut, we draw the ellipse on the ground.

Example 2: if φ = 46° and OA = 2 m, we obtain $OF_1 = OF_2 = 1.39$ m
There is no point in using greater distances than 2 meters for OA (and OB), since it would be impossible to read the time in summer if a person were to be used as a gnomon (Figure 8.4).

The hour points

Now that we have drawn our ellipse, we have to locate the hour points. There are various ways to do this.

Figure 8.3 The "gardeners' method". The ellipse is a series of points M such that the distance $F_1M + F_2M$ is constant. The length of the string is 2 OA.

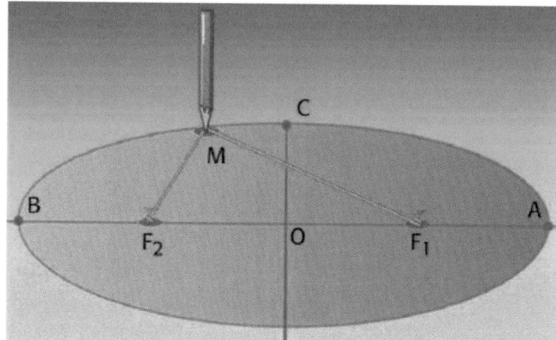

Figure 8.4
Analemmatic
sundial on the
Promenade du
Peyrou in
Montpellier
(Hérault). (SAF/
R.Sagot)

First method   We draw a circle of radius OA, with centre O, upon which we mark eight points, all 15° apart, to either side of noon, which is situated at C (Figure 8.5). Through these points we draw straight half-lines parallel to the north–south axis, i.e. at right angles to the east–west axis. Then we multiply the length of each semi-chord by the coefficient (sin φ). This gives the required hour point on the ellipse. We must not forget to mark all points from 4 h to 20 h, since this type of sundial will work from sunrise to sunset.

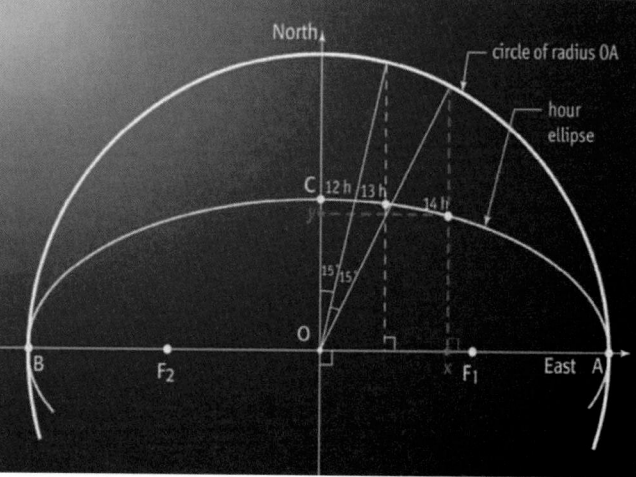

Figure 8.5
Positions of hour points. The hour points are obtained either by projection onto the ellipse of a circle graduated every 15°, or by calculating their coordinates.

Second method   To position the hour points along the ellipse, it is possible to calculate their x and y coordinates, with reference to point O (the x coordinates are reckoned along the semi-major axis positively eastwards, and the y coordinates are reckoned along the semi-minor axis positively northwards). The expressions for the coordinates are simple:

$$x = OA \sin H$$

$$y = OA \sin \varphi \cos H$$

where H is the hour angle of the Sun: $13\,h = 15°, 14\,h = 30°, \ldots, 19\,h = 105°$. H varies from $0°$ (noon) to $+120°$ (sunset), and from $0°$ to $-120°$ (sunrise). It is easy to see that the morning and afternoon points are symmetrical. The $6\,h$ and $18\,h$ points lie on the major axis of the dial. With the points marked on the ellipse, we can add floral motifs or decorative numbers.

Example: if $\varphi = 46°$ and $OA = 2\,m$, we obtain the hour points given in Table 8.1. For the morning points we have the same values of y; only the x values become negative.

The scale of dates
All that remains now is to draw the 'scale of dates' along which the style (which may well be a person) can move. The scale of dates is drawn along the north–south axis (Figure 8.6). Point O corresponds to the position of the style at the equinoxes, point E to that at the summer solstice and point H to that at the winter solstice. E is to the north of O and H is to the south of O. Distance $OE = OH = OA \cos \varphi \tan 23°.44$ ($23°.44$ being the angle of the obliquity of the

Table 8.1 The coordinates of hour points for $\varphi = 46°$ and $OA = 2\,m$

Solar time	Hour angle H	Coordinates of hour points
12 h	0°	$x = 0\,m$; $y = 1.44\,m$
13 h	15°	$x = 0.52\,m$; $y = 1.39\,m$
14 h	30°	$x = 1\,m$; $y = 1.25\,m$
15 h	45°	$x = 1.41\,m$; $y = 1.02\,m$
16 h	60°	$x = 1.73\,m$; $y = 0.72\,m$
17 h	75°	$x = 1.93\,m$; $y = 0.37\,m$
18 h	90°	$x = 2\,m$; $y = 0\,m$
19 h	105°	$x = 1.93\,m$; $y = -0.37\,m$
20 h	120°	$x = 1.73\,m$; $y = -0.72\,m$

ecliptic). If d refers to the position of the style with reference to O on a given day, it can be shown that:

$$d = OA \cos \varphi \tan \delta$$

where δ is the declination of the Sun on that day (see Appendix C, page 148). When $\delta > 0°$, the style is to the north of O; when $\delta < 0°$, it is to the south of O. Usually, dates corresponding to noteworthy declinations of the Sun are shown on the scale of dates (such as the Sun's entry, for example, into the various signs of the Zodiac: see Table 8.2.

Example: if $\varphi = 46°$ and $OA = 2\,m$, then $OE = OH = 0.6\,m$. The scale of dates is 1.2 m long. If $\delta = +20°.15$, then $d = 0.51\,m$. If $\delta = -11°.47$, then $d = -0.28\,m$.

Conclusion

Our sundial is now completed. To read the time, the person stands on the scale of dates at the appropriate position for the date of the observation. The time can be read at the point where his/her shadow meets the ellipse (Figure 8.7). Remember

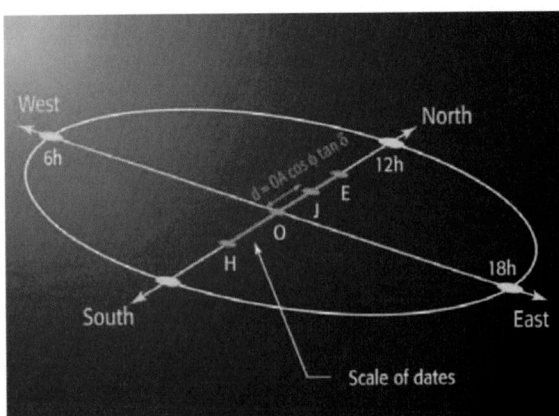

Figure 8.6 The scale of dates. The dates are marked on the meridian. A point on this scale indicates the place where the gnomon must be situated in order to tell the time on the corresponding date.

Table 8.2   Declination of the Sun as it enters the various signs of the Zodiac

Declination	Dates	Signs of Zodiac
$\delta = -23°.44$	21 December	Capricorn
$\delta = -20°.15$	20 January and 22 November	Aquarius and Sagittarius
$\delta = -11°.47$	19 February and 23 October	Pisces and Scorpio
$\delta = 0°$	20 March and 23 September	Aries and Libra
$\delta = +11°.47$	20 April and 23 August	Taurus and Virgo
$\delta = +20°.15$	21 May and 23 July	Gemini and Leo
$\delta = +23°.44$	21 June	Cancer

that it may be necessary to correct for the difference between solar time and local time. Of course, we cannot expect the sundial to be very accurate, given the width of the shadow, but greater accuracy can be achieved if we place on the scale of dates a movable gnomon which casts a much narrower shadow.

Figure 8.7 Analemmatic sundial at Gruissan (Aude). A person serves as a mobile gnomon. (S.Grégori

Lastly, it is worth noting that, in the case of an analemmatic sundial, the time is indicated by the direction of the shadow of the gnomon and not by the alignment of the shadow with an hour line as on other sundials. It should not be thought that the shadow of the gnomon describes an ellipse during the course of the day; the tip of the shadow moves along a hyperbolic path, but this is not important since here it is the direction of the shadow that counts. It is also worth mentioning that we do not need to mark out the whole of the hour ellipse, unless it is for aesthetic reasons. At the end of this book the reader will find a table giving values of x and y for different latitudes.

8.3 Demonstration of the principle of the analemmatic sundial

In specialized shops can be found embroidery frames (see section 4.5) consisting of two circular wooden hoops over which fabric is stretched while being worked upon. Here, we shall use one of these frames to demonstrate the principle of the analemmatic sundial, and to aid our understanding of two of its particular properties: why the hour points lie on an ellipse, and why the style has to move as a function of the date.

To this end we shall use the embroidery frame as an equatorial armillary sundial (cf. Chapter 4), and a length of electrical wire as a polar style.

Method

- **Materials required:** hoop from an embroidery frame; stiff electrical wire; sliding bead; small clothes-peg; black marker pen; protractor; wooden poles.

Having prepared our hoop by dividing it into sixteen $15°$ sectors using a pencil, on both upper and lower faces, we position the wire within it as near as possible to the central point and incline the system such that the wire points towards the North Celestial Pole. It is pointless to aim at accuracy with our 'rough and ready' armillary sundial, with its hoop representing the plane of the Equator and the wire serving as the central polar style. We slide a large bead onto the wire. The bead is held in position by a small clothes-peg. In an ideal version, the axis of the wire will make an angle with the horizontal equal to that of the local latitude, and the hoop an angle with the southern horizon of $(90° - \text{latitude})$. The plane of the hoop is now parallel to that of the Celestial Equator, and the lowest point on the hoop corresponds to noon (Figure 8.8).

On the day of the summer solstice, at solar noon exactly, we adjust the position of the bead on the axis so that its shadow falls exactly upon the noon point on the hoop. Throughout the day, the shadow of the bead will move successively across the sectors of the hoop. For example, one hour after noon the shadow will be at the 13 h point, and so on. We can verify that, on this day (the summer solstice), the straight line passing through the bead and its shadow makes an angle of about $23°.5$ (the declination of the Sun) with the plane of the hoop. In April, the bead

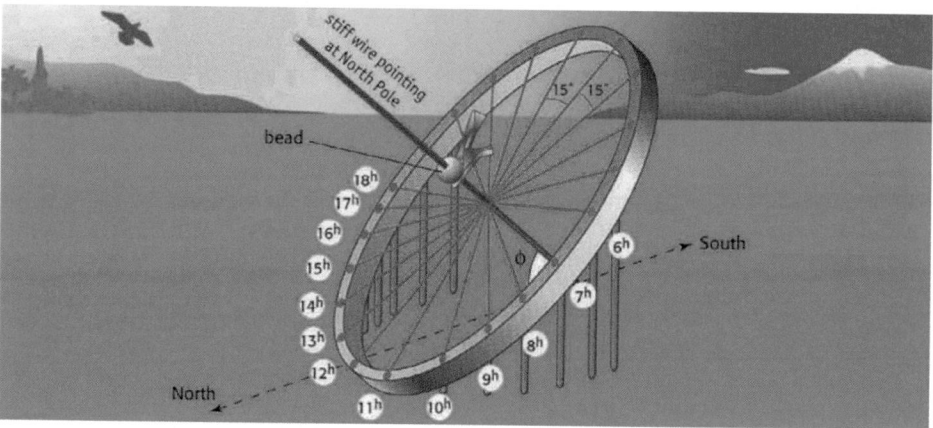

Figure 8.8 An equatorial armillary dial made from an embroidery hoop. A small vertical stick is placed beneath each hour mark on the hoop.

would to be set further down the axis; at the equinoxes it would lie in the plane of the hoop; and in winter, it would be on the southern section of the axis, below the plane of the hoop. This experiment clearly shows that it is necessary to move the bead along the axis as a function of the season, to ensure that its shadow falls in the right place on the hoop.

Let us now place a small vertical pole beneath each hour point. In France we will need 17 such poles: 8 for the points before noon, 8 for the points after noon, i.e. from 4 h to 20 h. Lastly we set up a vertical pole beneath the bead (it is still 21 June). What do we observe? At noon, the pole below the bead casts its shadow on the ground and on the noon point. At 13 h, the shadow lies on the ground and upon the 13 h pole; and so on. Consequently, the locations of the poles on the ground will suffice to tell the time. We mark these positions with our marker pen and take away the hoop. We then discover that the marks indicating where the poles stood lie on an ellipse. To read the time, we need to relocate the gnomon which carried the bead!

Further information:

- Drawing an analemmatic sundial using an equatorial sundial: see Appendix E, page 157.

8.4 Making a solar compass

Theory

With a solar compass we can determine geographical north by using the Sun. The simple construction suggested here consists of two conjoined sundials, one horizontal and the other analemmatic. When the two dials show the same time, the axis of the compass points north. Such a system will only work, however, for a given latitude: if we construct the compass for Le Mans, for example, it can be used only at latitude 48°. This is an obvious inconvenience, but the advantage of such a compass is that it indicates geographical north and not magnetic north.

Method

- **Materials required:** fairly thin board (A4 size); two small wooden slats; piece of thin wood from which the styles will be cut.

On the vertically placed board, we draw a straight line 12 cm from the bottom edge: this will be the dividing line between the horizontal and the analemmatic dials (Figure 8.9). 10.5 cm from the edges, we draw another straight line at right angles to the line already drawn. This second line runs the length of the board, and ends at the top of the board in an arrow, which will indicate north. Let us call the point where the two lines intersect point C. We mark point O, representing the centre of the analemmatic sundial, at a distance of 10 sin φ from point C. Point O will be towards the bottom of the board. The distance OC varies from 7.7 cm for latitude 50° to 6.8 cm for latitude 43°. The east-west axis passes through O. We then draw the hour points along the ellipse as indicated in section 8.2.

In the present case, we find the coordinates of the hour points with:

$$x = 10 \sin H$$

$$y = 10 \sin \varphi \cos H$$

The x axis corresponds to the east-west axis, and the y axis to the north–south axis.

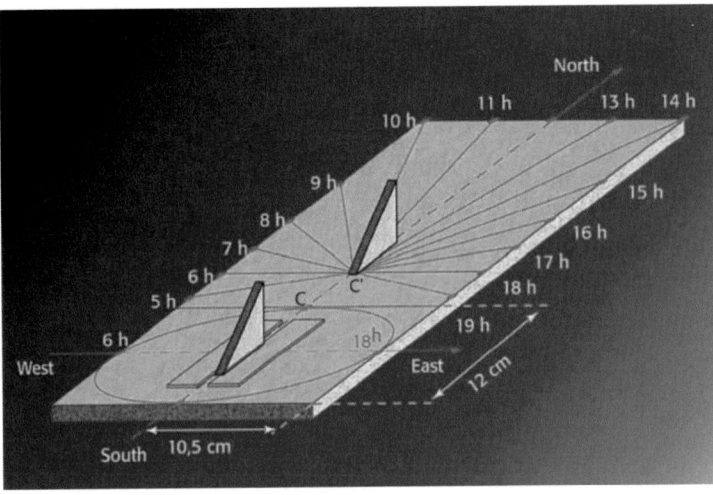

Figure 8.9
A solar compass. When the two dials show the same solar time, the axis of the two dials faces geographical north.

It remains only to draw the scale of dates from point O: north of O we use the quantity $10 \cos \varphi \tan(23°.44)$, and south of O, the quantity $-10 \cos \varphi \tan(23°.44)$.

The style of this dial can be in the form of a triangle 8 cm high, held upright by the two slats, on the scale of dates. Note that, in this case, it is the shadow of the vertical edge of the triangle, and not the shadow of its hypotenuse, that tells the time! Now, we create the horizontal sundial. The polar style will be placed on the north-south axis at C', 4 cm above point C. We draw a straight line through C', at right angles to the north–south axis. This is the 6 h–18 h line. Then we mark out the hour lines on the horizontal sundial by calculating the angle H' between the line and the noon line (which is the north-south axis) using the relationship:

$$\tan H' = \sin \varphi \tan H$$

Our chosen dimensions give us a fan of hour lines from 5 h to 19 h. For greater accuracy, we can work out the angles with reference to their tangents rather than by using a protractor (see page 74).

The polar style of the horizontal sundial must be triangular, such that the hypotenuse makes an angle with the plane of the dial equal to the local latitude. A style 6 cm high will suffice.

The compass

How do we use this compass? First of all, the Sun must be shining. We place the style of the analemmatic sundial in the position corresponding to the date. Then we orient the two sundials until they indicate the same time. This involves swinging the system horizontally until both dials' shadows are in agreement. The compass now indicates north, and also gives the solar time.

The analemmatic sundial at other latitudes

The coordinates of a point on the ellipse of the analemmatic sundial are, where a is the length of the semi-major axis:

$$x = a \sin H$$

$$y = a \sin \varphi \cos H$$

and the displacement d of the gnomon being

$$d = a \cos \varphi \tan \delta$$

• What happens to the analemmatic sundial on the Equator ($\varphi = 0°$)?
Answer: the ellipse becomes a straight line (since $y = 0$) along which lie the hour points.

• What happens to the analemmatic sundial at the North Pole ($\varphi = +90°$)?
Answer: the ellipse becomes a circle of radius equal to the semi-major axis, and the hour points are spaced 15° apart. As for the gnomon, it no longer moves! We have rediscovered the classic equatorial sundial.

Further information:
- Hour lines, analemmatic sundial: see Appendix B, page 145.

8.5 The circular analemmatic sundial

Theory

There are many different kinds of analemmatic sundial. Among these, the 'Lambert dial' is easy to construct. If a large-scale Lambert dial is to be attempted, it will no longer be possible to use a person as the style, since this analemmatic sundial features an inclined movable style. The Lambert dial is circular with evenly spaced (15°) hour lines, as in the case of the equatorial sundial. It is therefore one of the easiest to make (Figure 8.10).

Method

- **Materials required:** fairly thin board (A4 size); two small wooden slats; piece of wood from which the style will be cut.

We draw a circle with centre O and radius 15 cm. We then mark 17 evenly spaced points, 15° apart, on the circle: we can proceed as with the north face of an equatorial sundial. The hour points are marked from 4 h to 20 h, with the 12 h

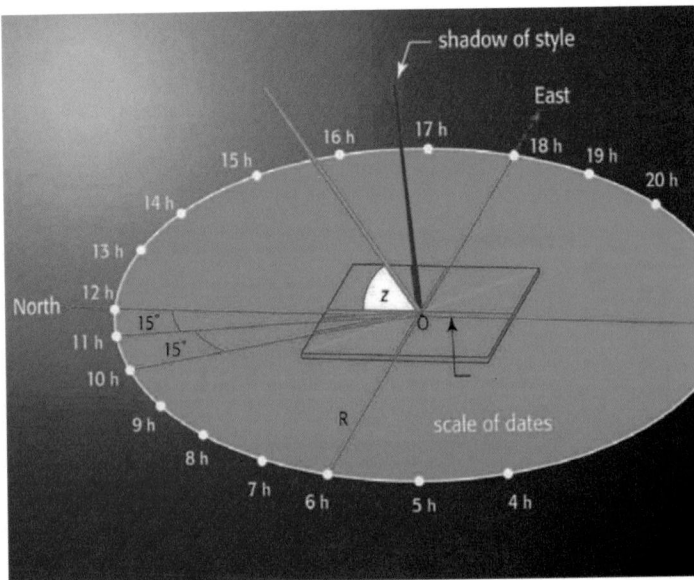

Figure 8.10
A circular analemmatic sundial. The style is at a constant angle, but it must be moved along the meridian as the year goes by. The hour points are on a circle, and are 15° apart.

point being on the meridian (towards geographical north of the style). As for the style itself, its inclination z to the horizon is equal to:

$$z = (90° + \varphi)/2.$$

The centre of the circle corresponds to the position of the inclined style at the equinoxes; the scale of dates, which coincides with the north-south noon line, must be graduated as with the elliptical analemmatic sundial. The displacement d of the style as a function of the date is given by:

$$d = R \tan z' \tan \delta, \qquad \text{where } z' = (90° - \varphi)/2$$

δ being the declination of the Sun (see Appendix C, page 148). If $\delta > 0°$, the points will be marked to the north of O; if $\delta < 0°$, the points will be marked to the south of O. The extreme points of the scale correspond to $\delta = \pm23° \ 26'$.

Example: if $\varphi = 48°$, $R = 15\,$cm and $\delta = \pm23° \ 26'$, then $z = 69°$, $z' = 21°$ and $d = 2.5\,$cm. Like the classic analemmatic sundial, the circular dial may be used in tandem with a horizontal sundial as a solar compass.

Local time

If we make the circle, with its hour lines, moveable around its axis, it is possible to indicate local time with this dial. We need only turn the circle until the style indicates the same time as a clock. For the rest of the day, the dial will indicate local time. Strictly speaking, this regulatory procedure will have to be repeated daily because of the variation of the equation of time.

9 Altitude sundials

Altitude sundials are amusing and original instruments. They tell solar time according to the altitude of the Sun above the horizon.

9.1 The principle of the altitude sundial

There are many different kinds of altitude sundial. They are all based on the same principle: the determination of solar time as a function of the altitude of the Sun (Figure 9.1). They also have the advantage of being portable, since they have to be moved to measure the altitude of the Sun. However, such dials also have their disadvantages: they are only usable for a given latitude, and are fairly unreliable between 11 h and 13 h solar time, as the altitude of the Sun varies little around noon. Also, it is necessary to know the date in order to use them. On the plus side, one of their advantages is that they can be used even if we do not know which direction is south.

Altitude sundials can take many forms (flat, cylindrical, ring-shaped, etc.). They usually consist of a system based on, for example, a hole, a style or a slit for measuring the Sun's altitude, a regulating system based on the date, and a network of hour lines upon which the time is indicated.

9.2 The Saint Rigaud Capuchin sundial

The theory of the Capuchin sundial

This sundial, invented in the seventeenth century by Father Saint Rigaud, takes its name from the resemblance of its markings to the hood worn by Capuchin monks. The dial is held vertically and turned in such a way that the Sun shines at a grazing angle across its face. The marker thread moves along a scale of dates, and, once the dial is set in the correct position *vis-à-vis* the Sun, the vertical shadow cast by the thread will indicate the solar time on the table of hours, using a sliding bead (Figure 9.2).

Method

- **Materials required:** wooden board (A4 size); length of fishing line and small lead weight; bead to run along the fishing line; small wooden or metal rectangle, with central slit; metal rule; protractor; compasses.

Figure 9.1 Cylindrical altitude sundial (shepherd's dial) in the park of the town hall in Châteaubernard (Charente). The shadow of the style, which moves as a function of the date, is projected on the cylinder, where solar time is indicated. (S.Grégori)

We take a lightweight piece of wood cut to the exact size of a sheet of A4 paper. Then we draw a straight line, exactly parallel to the upper edge of this board and 1 cm away from it. Then, at the right-hand end of this line and at right angles to the dial, we fix the small rectangle with its central slit. The slit should be 2–3 mm wide. This will constitute the alignment system. The dial must be held vertically such that the sunlight passing through the slit falls along the line. The following diagram must then be completed using a pencil. We draw a 180° semi-circle of radius OM, and the straight line MM′ (with O at its mid-point) parallel to the upper edge of the dial (Figure 9.3). The resulting shape is divided into 15° segments, with the noon point ($H = 0°$) at point M and the perpendicular passing through O at $H = 90°$ (6 h and 18 h). We can use a protractor to draw the dividing lines, or measure lengths (OM cos H) along MM′ from O, from which points we drop perpendiculars to the hour points corresponding to H. If the result is negative, the point in question will

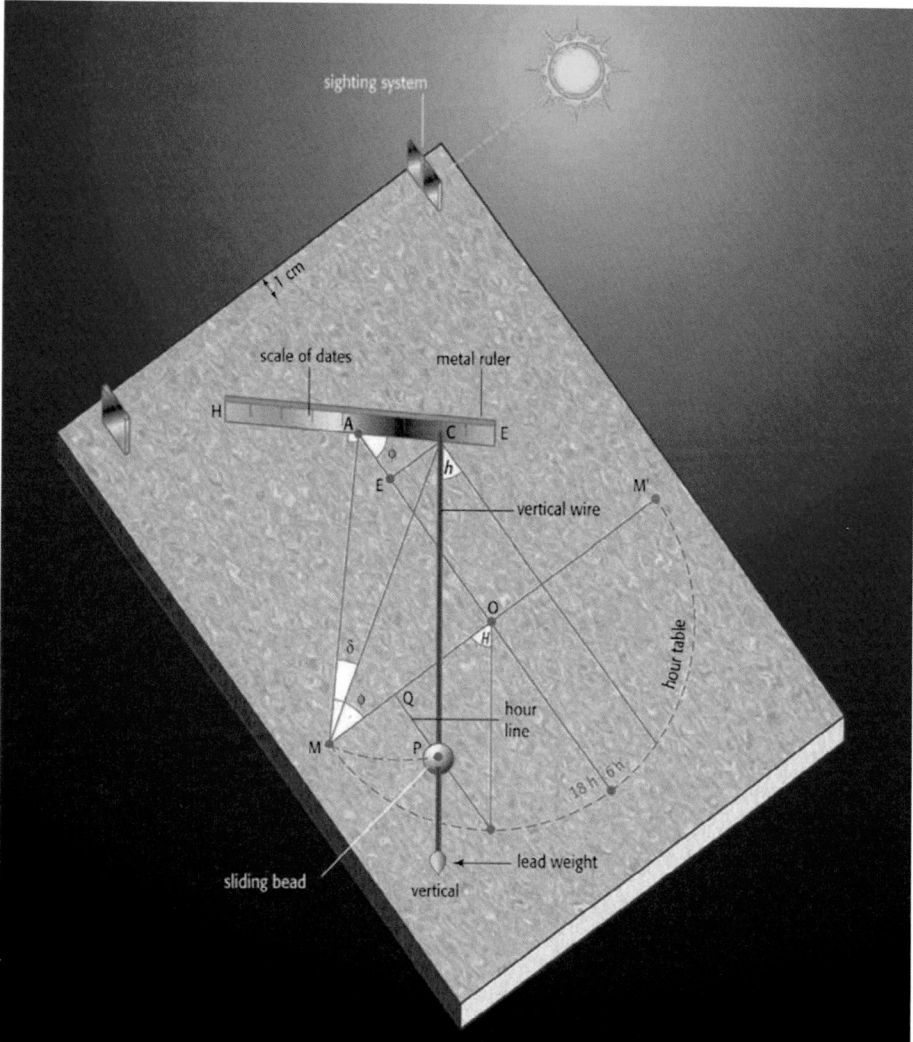

Figure 9.2 The Saint Rigaud Capuchin sundial. The sighting system allows the dial to be aligned with the vertical plane in which the Sun lies. The movable vertical thread on the scale of dates indicates the hour shown on the table.

lie to the right of O. Once we have established our hour points, we draw the hour lines from them to meet MM′ at right angles.

At our latitude, the dial is marked out for times between 4 h and 20 h. We can perform a 'double' graduation, since, by definition, point $H = 15° = 13$ h is the

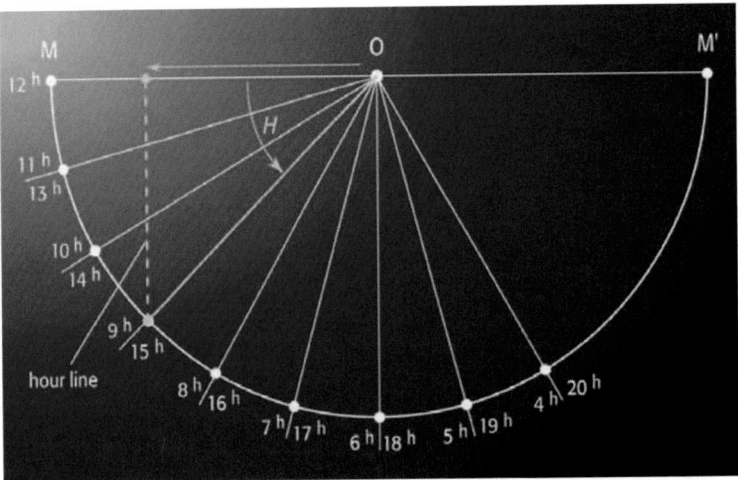

Figure 9.3 Drawing the table of hours. The semi-circle of radius OM is divided into 15° sectors. Each point (with the exception of the noon point) corresponds to two different hours, one in the morning and one in the afternoon.

same as that corresponding to $H = -15° = 11\,\text{h}$: the points are symmetrical, the altitude of the Sun for $H = 11\,\text{h}$ being the same as for $H = 13\,\text{h}$.

From M we draw a straight line MA, of length $\text{MA} = \text{OM}/\cos \varphi$. This line makes an angle with MM′ equal to the local latitude φ, such that $\text{OA} = \text{MA} \sin \varphi$ (Figure 9.4). Point A will therefore lie on the vertical line passing through O. At right angles to MA, we draw the scale of dates EH, along which will move our plumb-line, with its sliding bead. Points E and H mark the positions of the end of this plumb-line at the summer and winter solstices respectively. The distance $\text{AE} = \text{AH} = \text{OM} \tan(23°.44)/\cos \varphi$. Point C on the scale of dates, corresponding to a date in the year when the declination of the Sun is δ, lies at a distance from point A of $\text{OM} \tan \delta/\cos \varphi$. So, using the declinations of the Sun as listed in Appendix C, page 148, we mark as best we can the dates for the 1st and the 15th of each month.

Once we have completed our diagram, we can simply leave the 180° sector with its numbered hour lines, and erase the rest.

Example
There follows a detailed plan for drawing a Capuchin dial face for $\varphi = \text{latitude } 48°$. On a sheet of A4 paper, we draw a straight line parallel to the lower edge and 12 cm from it. Point O is marked at a distance of 10.5 cm from either edge. Let OM be 10 cm; MM′ will therefore be 20 cm (Figure 9.5). With a pair of compasses, we draw the 180° sector, and then we use a protractor to divide it, from point M, into 15°

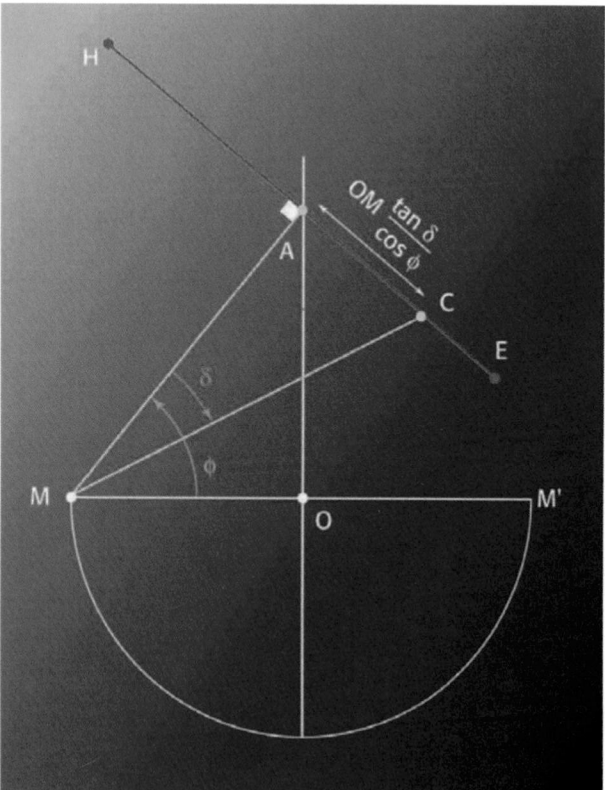

Figure 9.4 Drawing the scale of dates. On the segment EH are marked different dates in the year.

sectors up to and including 120°. From each point, we draw a perpendicular to MM′. If we use trigonometry, then point $11\,h = 13\,h = 15°$ will be 9.7 cm from O, and point $14\,h = 10\,h = 30°$ will be 8.7 cm from O (etc.). We now draw the scale of dates. Using a protractor, we draw a straight line MA, 14.9 cm long, at an angle of 48° to MM′. We can also prolong the straight line passing through O and intersecting MM′ at a right angle: in this case, OA will be 11.1 cm long.

Having established the position of point A, we draw a perpendicular to MA through this point. The distance $AE = AH = 6.5$ cm, which gives the limits of the scale. Let us now suppose that we wish to locate the point corresponding to May 1. From the table of declinations, we know that on that date $\delta = +15°\ 12'$. Therefore, this point will be 4.1 cm to the right of A. For the point corresponding to November 15, $\delta = -18°\ 35'$. Now, the point will be 5 cm to the left of A.

All that remains to do is to indicate the hours on our 180° sector, marking for example the morning hours above the line and the afternoon hours below it. We

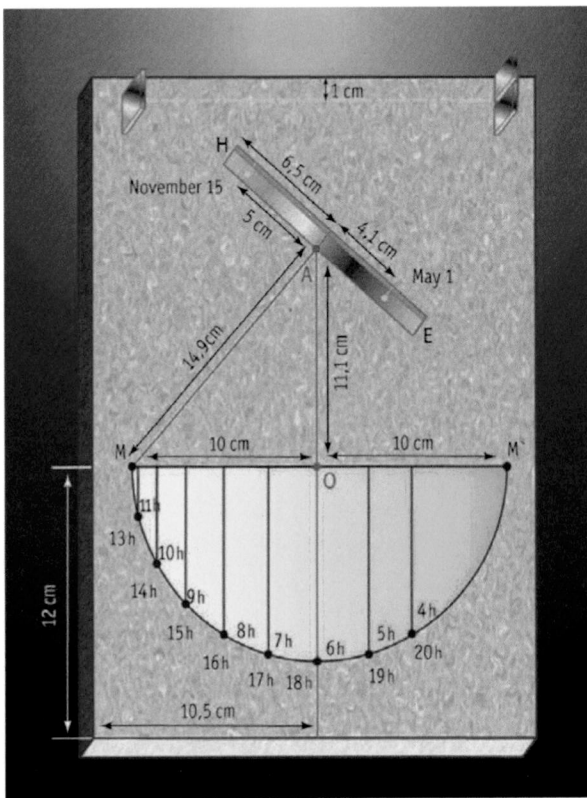

Figure 9.5 An example of the complete diagram on an A4-sized board.

will of course have drawn a straight line parallel to MM′ on the A4 sheet, 1 cm from its upper edge; this serves to orient the dial in the plane of the Sun.

The dial in use

Let us suppose that the plumb-line is placed at C, i.e. for a positive declination of the Sun (Figure 9.6). With the dial held vertically, we turn it such that the narrow beam of light falls on the metal screen mounted on the dial. The Capuchin is now in the vertical plane on which the Sun lies. We will have previously placed the sliding bead at P, such that CM = CP. With the plumb-line vertical, the bead at P will lie on one of the hour lines, and we can read the solar time.

Further information:

● Demonstration of the Saint Rigaud Capuchin sundial: see Appendix E, page 158.

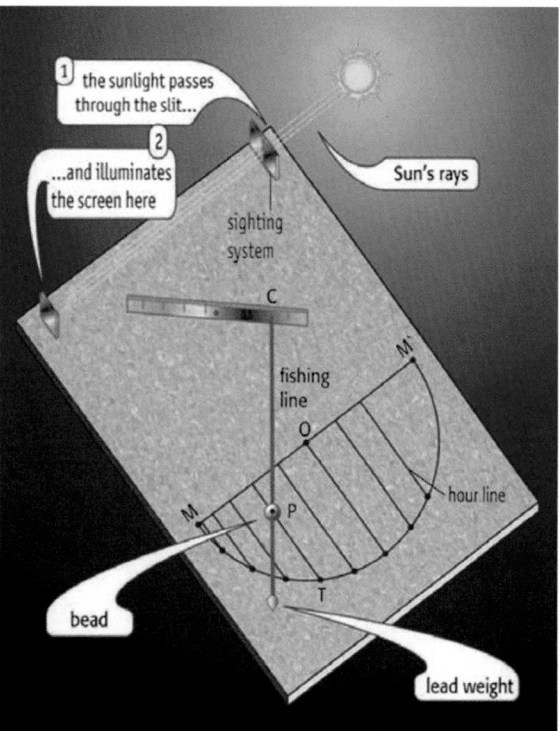

Figure 9.6 How a Saint Rigaud Capuchin sundial works. Once the thread has been correctly placed on the scale of dates, the sliding bead is adjusted such that CP = CM. Then, the sighting system is aligned with the Sun.

10 Sundials in the tropics

The apparent motion of the Sun in intertropical zones is very different from that observed in temperate latitudes. Making a sundial for these zones can lead us down some surprising and compelling paths.

10.1 Sundials in the tropics

For those living in tropical zones, the procedures described above for making sundials will not be the same. Sadly, books on the subject rarely mention this aspect. In these regions, it is possible to carry out a unique experiment with a simple gnomon set in the ground, that of studying the retrograde motion of the shadow (see Appendix F, page 163). We shall examine two scenarios: that of a sundial at a location between the Tropic of Cancer and the Equator ($+23°$ $26' > \varphi > 0°$), and that of a sundial at a location between the Tropic of Capricorn and the Equator ($-23° 26' < \varphi < 0°$).

10.2 Sundials north of the Equator

The apparent motion of the Sun
Remember that, between the Equator and the Tropic of Cancer, the Sun will culminate in the south, as long as the value for the declination of the Sun is less than that of the local latitude. The Sun will pass through the zenith when its declination is equal to the local latitude. Then, when the value for declination of the Sun is greater than that of the local latitude, the Sun will culminate in the north (Figure 10.1). In other words, the Sun passes through the zenith twice a year for locations between the Equator and the Tropic of Cancer. For a location exactly on the Tropic of Cancer, it passes through the zenith only once a year, on 21 June.

Also, the daylight period in tropical zones remains much the same (about 12 hours) all year long: the Sun rises at around 6 h and sets at around 18 h. On the Tropic itself, this period is longest: on 21 June the Sun rises at 5 h 17 min and sets at 18 h 43 min.

The gnomon
Finding the local meridian using a gnomon should be done as far as possible when the value for the declination of the Sun is at its furthest from that of the latitude. For example, at a location at latitude $+14° 31'$, it is best to do this in December: the

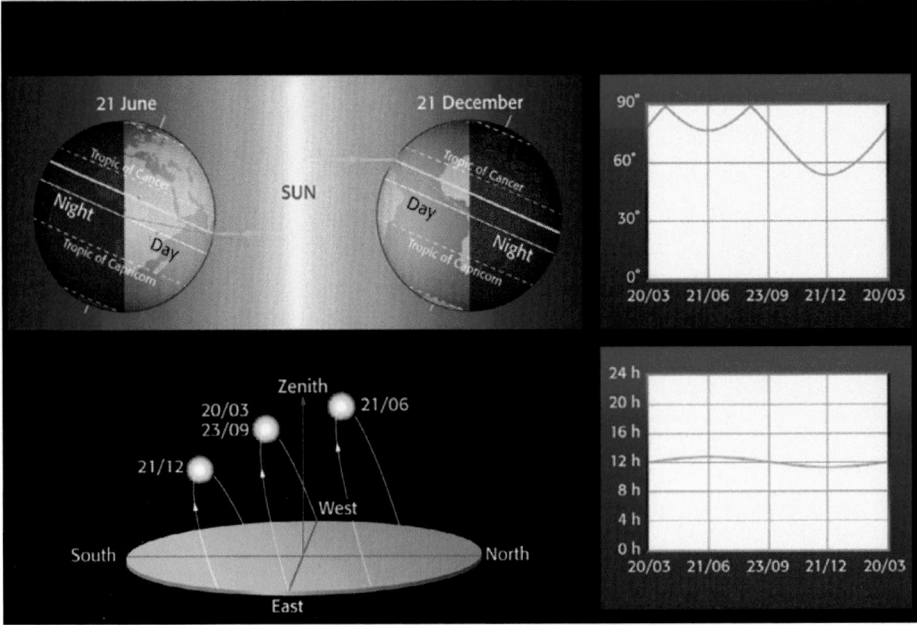

Figure 10.1 Between the Tropic of Cancer and the Equator, the Sun passes through the local zenith twice a year, as shown in the panel top right.

shadow is at its longest at noon. Note that, at the moment of the Sun's meridian passage, a miscalculation of just a few seconds at the time of marking the position of the shadow of the gnomon will lead to serious errors: the azimuth of the Sun changes very rapidly when it is near the zenith.

If we mark out a solar calendar, for example in a school playground, with a gnomon, it will be of interest to satisfy ourselves that the dates indicated indeed lie on both sides (north and south) of the gnomon. If we mark out a seasonal indicator in the Caribbean, for example, we will see that the hyperbola representing the summer solstice curves to the south of the foot of the gnomon. On the Equator, the hyperbolic arcs are symmetrical on either side of the straight line representing the equinoxes, on which line the foot of the gnomon is situated (Figure 10.2).

It is worth pointing out that the altitude h of the Sun at solar noon is calculated differently according to whether the value for the Sun's declination δ is smaller or greater than the local latitude φ. In the first case, $h = 90° - \varphi + \delta$; in the second, $h = 90° + \varphi - \delta$. For example, if $\varphi = 15°$ and $\delta = 2°$, then $h = 77°$ (the Sun culminates in the south). If $\varphi = 15°$ and $\delta = 23° \ 26'$, then $h = 81° \ 34'$ (the Sun culminates in the north). On the Equator, the altitude of the Sun varies between $(90° \pm 23° \ 26')$.

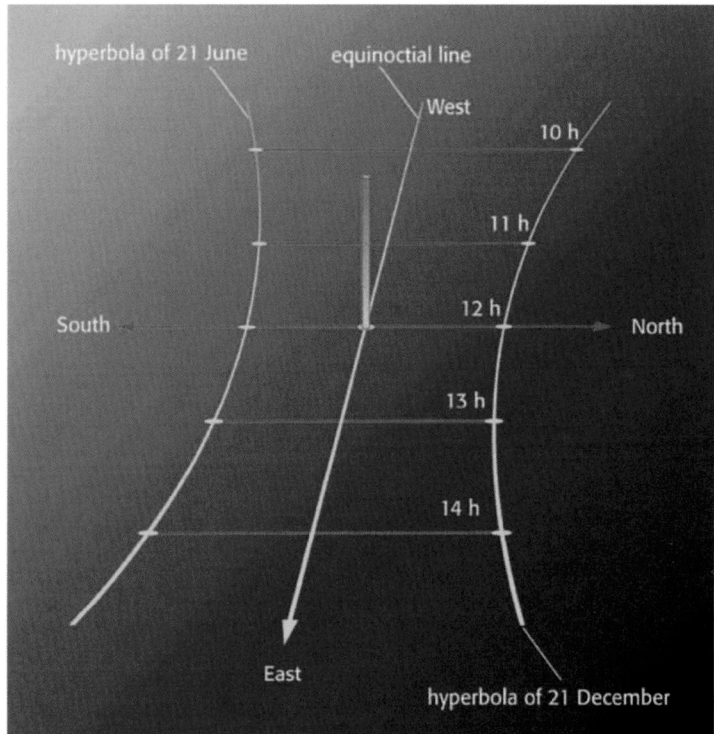

Figure 10.2 A solar calendar with a gnomon at the Equator. The diurnal arcs of the calendar lie to either side of the east-west axis.

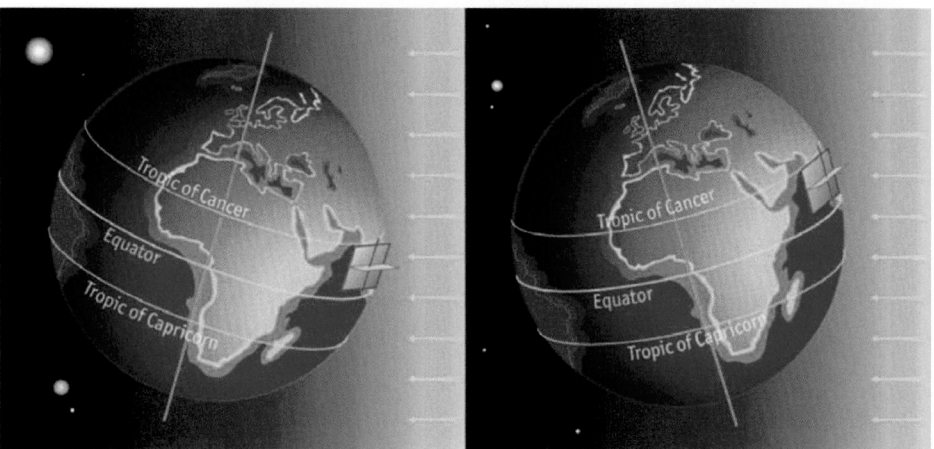

Figure 10.3 An equatorial sundial between the Tropic of Cancer and the Equator. The north face of the dial is illuminated only when the declination of the Sun is greater than the local latitude.

The equatorial sundial

At these latitudes, the equatorial sundial assumes a new aspect: the table is not much inclined towards the horizon, and at the Equator itself it becomes vertical. As in the case of an equatorial sundial drawn for a European latitude, it is intriguing to note that the shadow of the style travels anti-clockwise as we look at the dial while facing north, and in a clockwise direction, if we look at it while facing south.

The horizontal sundial

The polar style of the horizontal sundial is not much inclined towards the northern horizon. The style must be very long, in order to be able to read certain times of the day when the Sun is very high in the sky. As with the equatorial sundial, the outer hour lines are around 6 h and 18 h. The distribution of the hour lines is very different from that seen in latitudes further north: the lines representing 9 h, 10 h, 11 h, 13 h, 14 h and 15 h are quite close to the noon line, while the other lines are much more widely spaced (Figure 10.4). Table 10.1 gives, as an example, the angle between the afternoon hour lines and the noon line for latitude +16°.

If we are exactly on the Equator, the diagram is identical to that for the polar sundial: the lines are all parallel and the style itself must be parallel to the dial table. What is more, the closer we are to the Equator, the more the polar sundial is to be preferred to the horizontal sundial. Indeed, marking out a classic horizontal

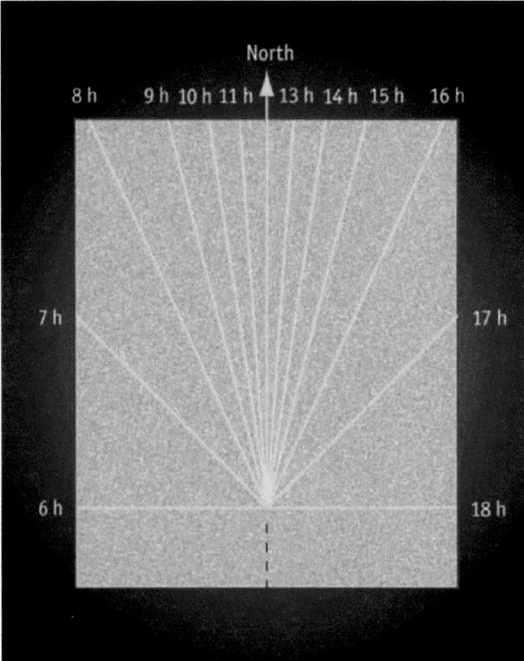

Figure 10.4 Hour lines on a horizontal dial at latitude +16°.

sundial at latitude 5° is problematical, since the lines are very close together around noon and actually reading the time becomes quite difficult. Table 10.2 gives, as an example, the angle between the afternoon hour lines and the noon line for latitude +5°.

The vertical sundial

Vertical sundials pose no particular problems. The direct south meridional dial still has its hour lines limited to between 6 h and 18 h. Also, its diagram is quite close to that of the equatorial sundial, as the angles between the hour lines are almost 15°. Moreover, the polar style is practically horizontal. However, depending on the season, it will not always work at noon: when the value for the declination of the Sun is greater than that of the local latitude, noon is indicated on the north side. Now, as the Sun climbs very high in the sky, the shadow of a style on the sundial is very long, and it is sometimes difficult or even impossible to mark the hyperbolic arcs indicating the seasons.

Making the diagram for a vertical direct north sundial is interesting, because it is no longer limited, as in Europe, to just a few lines in the morning and late afternoon. All the hours from 6 h to 18 h must be drawn: the septentrional dial can indicate noon, all the while the value for the declination of the Sun is greater than that of the local latitude.

If we are exactly on the Equator, the north face will be in use for six months (20 March—23 September), and the south face for the next six months (23 September—20 March). The style of each dial is here perpendicular to the wall.

Table 10.1 Hour line angles H' from the noon line for a horizontal dial at latitude +16°

Solar time	H'
12 h	0°
13 h	4°.22
14 h	9°.04
15 h	15°.41
16 h	25°.52
17 h	45°.81
18 h	90°

Table 10.2 Hour line angles H' from the noon line for a horizontal dial at latitude +5°

Solar time	H'
12 h	0°
13 h	1°.34
14 h	2°.88
15 h	4°.98
16 h	8°.58
17 h	18°.02
18 h	90°

The analemmatic sundial

Marking out an analemmatic sundial presents a few problems in the tropical zones. First of all, the ellipse of the sundial is very flattened, becoming a straight line at the Equator. The scale of dates is longer than the minor axis of the ellipse (except at the Tropic itself, where they overlap). So, on 21 June, a person acting as a gnomon stands outside the ellipse. This is not too problematical, but, since the time is indicated by the direction of the shadow (which must cut the ellipse), the person has to be quite tall for the shadow actually to reach the noon point! Nevertheless, this restriction should not discourage us from marking out analemmatic sundials! We could opt for more modest dimensions, for example those of an A4 or A3 sheet.

10.3 Sundials south of the Equator

The apparent motion of the Sun

In the southern hemisphere, latitudes have negative values. Remember that between the Equator and the Tropic of Capricorn, while the value for the declination of the Sun is less than that of the local latitude, the Sun will culminate in the south. The Sun will pass through the zenith when its declination is equal to the local latitude. Then, when the declination of the Sun is greater than the local latitude, the Sun will culminate in the north (Figure 10.5). For locations south of the Tropic of Capricorn, the Sun always culminates in the north.

We can still make use of the equation involving the altitude of the Sun at solar noon, as long as we take into account the sign of the value of the latitude. For example, at $\varphi = -17°$ on December 21 ($\delta = -23°\ 26'$), the altitude h of the Sun is equal to $90° - \varphi + \delta$, i.e. $83°\ 34'$ (the Sun culminates in the south). If the Sun culminates in the north, and the value for its declination is therefore smaller than that of the local latitude, we use $90° + \varphi - \delta$. So on June 21 ($\delta = +23°\ 26'$), the altitude of the Sun is $49°\ 34'$.

Sundials

The styles of sundials located in the southern hemisphere point towards the South Celestial Pole; at night, it becomes obvious that there is no Pole Star worthy of the name at that point in the sky. The remarks in the previous section on the use of gnomons are still valid. In Tahiti (latitude $-17°\ 40'$), for example, it is best to trace out the local meridian when the declination of the Sun is around $+23°\ 26'$, i.e. in

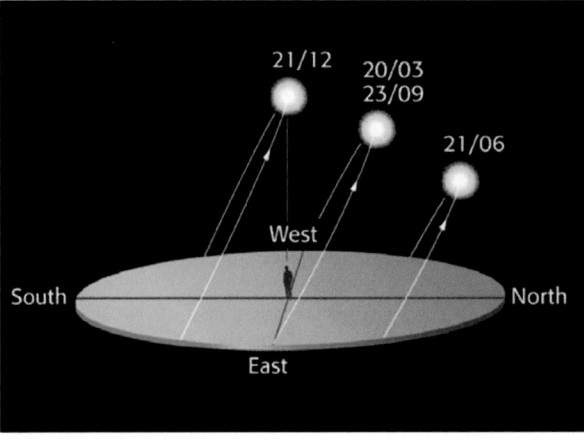

Figure 10.5 Between the Equator and the Tropic of Capricorn, the Sun passes through the local zenith twice a year.

June. The horizontal sundial in the southern hemisphere has the particular feature that the morning hours are shown on the right-hand (western) side, and the afternoon hours are shown on the left-hand (eastern) side. The shadow of the polar style, which makes an angle with the horizon equal to the local latitude, points due south at noon. As for vertical sundials, it is the northern face that is most often illuminated, especially in locations near to the Tropic of Capricorn. The shadow of the style travels in a clockwise direction, and in the opposite direction on the southern face. As is always the case in the tropics, the noon line is on both faces. On the island of Réunion, for example, at latitude $-21°$ $12'$, the southern face is illuminated from 27 November until 15 January; for the rest of the year, it is the northern face which indicates the time.

Further information:
- Under what circumstances can we observe the retrograde motion of the shadow of a gnomon? see Appendix F, page 163.

Appendix A

A little test

1 A sundial in Moscow (longitude 37° 30′ east) shows solar noon. At the same moment, what time is shown on a sundial on the Greenwich Meridian?

2 On 20 March at solar noon, a vertical gnomon has no shadow. What is the local latitude?

3 The height of the Eiffel Tower is 319 meters. How long is its shadow at solar noon at the winter solstice? (The latitude of Paris is 48° 50′)

4 The Great Pyramid of Cheops on the Gizeh Plateau near Cairo (latitude 30°) is 137 meters high, and the sides of its square base are 230 meters long. Given that its four faces are oriented towards the cardinal points, will the pyramid cast a shadow on the ground on 21 June at solar noon?

5 Both sides of a wall running east–west and located on the Tropic of Capricorn are sunlit. What is the date?

6 Where on planet Earth can a horizontal sundial indicate midnight?

7 As the Sun rises, a sundial attached to a wall reads 6 h. On which two dates is this possible?

8 In France, the Greenwich Meridian passes through the Cirque de Gavarnie in the Pyrenees. At what (clock) time will the Sun cross the meridian on Christmas Day at this location?

9 The time for the Muslim ʹasr daily prayer is defined thus: when a man's shadow is equal in length to his shadow at solar noon, plus his own height. Express mathematically the length of the shadow at this time.

10 Why was the shadow of a gnomon at noon shorter at the summer solstice in the time of Pericles (about 450 BC) than it is today?

11 At true noon, on the day of the winter solstice (21 December), the shadow of a gnomon is twice the length of the gnomon. What is the local latitude?

12 Some travelers leave France and are shipwrecked on a desert island sometime between 21 December and Christmas Day. They still have a watch and a calculator. They erect a gnomon on the beach and observe its shadow (which points north) when the shadow is at its shortest; they come to the conclusion that the shadow is exactly half the length of the gnomon itself when it is 8 h 07 by the watch. Where are they, approximately (latitude and longitude)?

Answers to test

1   Expressing the longitude of Moscow in hours, minutes and seconds by dividing the longitude (in degrees) by 15, we obtain −2 h 30 min. This means that when the Sun crosses the meridian in Moscow, i.e. when it is solar noon, it is 2 h 30 min earlier on the Greenwich meridian, i.e. 9 h 30 min solar time.

2   If a gnomon has no shadow at solar noon, the Sun must be at the zenith for that location. This can only happen within the tropical zones. As 20 March is the day of the spring equinox in the northern hemisphere; in other words, the Sun crosses the Celestial Equator. It is therefore at the zenith on the terrestrial Equator, and the latitude in question is $0°$.

3   We know that, at solar noon, the Sun culminates on the meridian. Its altitude h is then $90° - \varphi + \delta$, where φ is the local latitude and δ the declination of the Sun. The latitude of Paris is $48° \ 50'$; on the day of the winter solstice, the declination of the Sun is $-23° \ 26'$. Therefore, the altitude of the Sun on the meridian is equal to $17° \ 44'$. If a is the height of the Eiffel Tower and l the length of its shadow, then $a/l = \tan h$, whence $l = a/\tan h$. Therefore the length of the shadow is 997.6 meters, or nearly a kilometer! In reality, a considerable penumbral effect means that no sharp shadow of the top of the Tower can be distinguished.

4   On 21 June at the latitude of Cairo, the Sun's altitude h at solar noon is equal to $90° - \varphi + \delta$, i.e. $90° - 30° + 23° \ 26' = 83° \ 26'$. Let us imagine that the pyramid is transparent, and that 137 meters represents the height of an imaginary gnomon, with a north-pointing shadow at noon on 21 June measuring $(137/\tan(83° \ 26')) = 15.8$ meters. Given that the base of the pyramid has sides 230 meters long (the distance between the foot of the 'gnomon' and the northern edge of the base being 115 meters), it is obvious that the shadow is far too short to reach the edge of the pyramid. Let us now consider the actual pyramid. The inclination of the slope on the northern face is equal to arctan $(137/115) = 49° \ 59'$. Since the Sun is $33° \ 27'$ higher, it illuminates the northern face (and the southern face). Therefore, there is no shadow on the ground, nor is there a shadow on the pyramid!

5   The latitude of the Tropic of Capricorn is $-23° \ 26'$. If the wall in question, with sides facing north and south, is illuminated at solar noon, this means that the Sun is at the zenith, and the date is therefore 21 December.

6   For the Sun to be above the horizon at true midnight, we must be in the polar zones: in the northern hemisphere, this can happen between the Arctic Circle (latitude $+66° \ 34'$) and the North Pole (latitude $+90°$). In the southern hemisphere, this can happen between the Antarctic Circle (latitude $-66° \ 34'$) and the South Pole (latitude $-90°$).

7   If the sundial shows 6 h at sunrise, this means that the Sun is rising exactly in the plane of the wall. Now, the Sun rises at 6 h solar time only twice a year, at the equinoxes (20 March and 23 September).

8 The longitude of the Cirque de Gavarnie is $0°$, since it is on the Greenwich Meridian. If we set up a sundial there, we can convert the solar time (as shown by the sundial) to clock time by adding the equation of time +1 hour, since, at Christmas, Winter Time is in force. Since the equation of time is zero on 25 December, when the sundial shows noon, the clock will show 13 h.

9 Let us call a man's height a and the length of his shadow (at solar noon) l. Then $l = a/\tan h$, where $h = 90° - \varphi + \delta$, i.e. $l = a/\tan(\varphi - \delta)$. Therefore the length of the shadow at the time for the 'asr prayer is $(a + l)$ or $[a + a \tan(\varphi - \delta)]$ or again $a(1 + \tan(\varphi - \delta))$.

10 The length l of the shadow of a gnomon a (or a stick) at solar noon is equal to $a/\tan h$, where h is the altitude of the Sun $(90° - \varphi + 23° \, 26')$ at the summer solstice. The quantity $23° \, 26'$ represents the maximum declination of the Sun, but it also represents the current value of the obliquity of the ecliptic. Now, this value is slowly decreasing, by about $1'$ per century (more accurately, $47''$ per century). More than 24.5 centuries have passed since the time of Pericles, when the value of the obliquity was $23° \, 26' + (24.5 \times 47'')$, i.e. $23° \, 45'$. Therefore, in Pericles' day, the Sun was higher in the sky at noon: so the shadow of a gnomon was shorter.

11 The length l of the shadow of a gnomon a (or a stick) at solar noon is equal to $a/\tan h$, where h is the altitude of the Sun $(90° - \varphi - 23° \, 26')$ at the winter solstice. If $l = 2a$, then $\tan h = 0.5$, whence $h = 26° \, 33' \, 54''$. From this we can deduce that the latitude φ is equal to $90° - h - 23° \, 26'$, or practically $40°$.

12 The travelers are marooned sometime between 21 December and Christmas Day, at a time of year when the declination of the Sun varies little, and is to all intents and purposes equal to $-23° \, 26'$. The fact that the shadow of their gnomon points north at local noon means that they cannot be south of the Tropic of Capricorn (or the shadow would point south).

The observation of the shadow at its shortest (i.e. when the Sun crosses the meridian of the island) reveals that it is exactly half the length of the gnomon. If a is the length of the gnomon and l the length of its shadow, then $l = a/2$. Now, the length of the gnomon divided by the length of its shadow gives the tangent of the Sun's altitude. Therefore, we have $\tan h = 2$, whence $h = 63° \, 26'$. Knowing that the Sun's altitude h at solar noon is equal to $90° - \varphi + \delta$, where φ is the local latitude and δ the declination of the Sun, we calculate the latitude from $\varphi = 90° - h + \delta$, i.e. $3° \, 08'$. Our travelers are just north of the Equator. The shadow is at its shortest at 8 h 07 min by their watch, which shows French time. Now, Winter Time being in force in France in December, and so we subtract one hour to find Universal Time: 7h 07 min UT. Since the equation of time is close to zero around Christmas, if it is solar noon on the island and 7h 07 min UT, the longitude of the island must be $7 \text{ h } 07 \text{ min UT} - 12 \text{ h} = -4 \text{ h } 53 \text{ min}$, or $73° \, 15'$ E. The castaways are therefore somewhere in the Maldives.

Hour lines

The tables below give, for vertical, horizontal and analemmatic sundials the position of the hour lines for different latitudes in mainland France.

Vertical sundials

Hour line		Angles from the noon line as a function of latitude									
morning	afternoon	42°	43°	44°	45°	46°	47°	48°	49°	50°	51°
11h	13h	11°.3	11°.1	10°.9	10°.7	10°.5	10°.4	10°.2	10°	9°.8	9°.6
10h	14h	23°.2	22°.9	22°.6	22°.2	21°.9	21°.5	21°.1	20°.7	20°.4	20°
9h	15h	36°.6	36°.2	35°.7	35°.3	34°.8	34°.3	33°.8	33°.3	32°.7	32°.2
8h	16h	52°.2	51°.7	51°.2	50°.8	50°.3	49°.8	49°.2	48°.7	48°.1	47°.5
7h	17h	70°.2	69°.9	69°.6	69°.2	68°.9	68°.6	68°.2	67°.8	67°.4	66°.9
6h	18h	90°	90°	90°	90°	90°	90°	90°	90°	90°	90°

Horizontal sundials

Hour line		Angles from the noon line as a function of latitude									
morning	afternoon	42°	43°	44°	45°	46°	47°	48°	49°	50°	51°
11h	13h	10°.2	10°.4	10°.5	10°.7	10°.9	11°.1	11°.3	11°.4	11°.6	11°.8
10h	14h	21°.1	21°.5	21°.9	22°.2	22°.6	22°.9	23°.2	23°.5	23°.9	24°.2
9h	15h	33°.8	34°.3	34°.8	35°.3	35°.7	36°.2	36°.6	37°	37°.5	37°.9
8h	16h	49°.2	49°.8	50°.3	50°.8	51°.2	51°.7	52°.2	52°.6	53°	53°.4
7h	17h	68°.2	68°.6	68°.9	69°.2	69°.6	69°.9	70°.2	70°.5	70°.7	71°
6h	18h	90°	90°	90°	90°	90°	90°	90°	90°	90°	90°
5h	19h	111°.8	111°.4	111°.1	110°.8	110°.4	110°.1	109°.8	109°.5	109°.3	109°
4h	20h	—	—	—	—	—	—	127°.8	127°.4	127°	126°.6

Analemmatic sundials

Using the x and y coordinates given below, we can locate the hour points on the ellipse of an analemmatic sundial for different latitudes. They are calculated for a semi-major axis $a(=1)$. The value of the semi-minor axis corresponds to the value of y at 12 h. If we take as an example $a = 150$ cm, all we need to do is multiply all the values of x and y by 150.

Hour point		Coordinates of hour point as a function of latitude									
morning	afternoon	42°	43°	44°	45°	46°	47°	48°	49°	50°	51°
12 h		$x = 0$	$x = 0$	$x = 0$	$x = 0$	$x = 0$	$x = 0$	$x = 0$	$x = 0$	$x = 0$	$x = 0$
		$y = 0.669$	$y = 0.682$	$y = 0.695$	$y = 0.707$	$y = 0.719$	$y = 0.731$	$y = 0.743$	$y = 0.755$	$y = 0.766$	$y = 0.777$
11 h	13 h	$x = 0.259$	$x = 0.259$	$x = 0.259$	$x = 0.259$	$x = 0.259$	$x = 0.259$	$x = 0.259$	$x = 0.259$	$x = 0.259$	$x = 0.259$
		$y = 0.646$	$y = 0.659$	$y = 0.671$	$y = 0.683$	$y = 0.695$	$y = 0.706$	$y = 0.718$	$y = 0.729$	$y = 0.740$	$y = 0.751$
10 h	14 h	$x = 0.500$	$x = 0.500$	$x = 0.500$	$x = 0.500$	$x = 0.500$	$x = 0.500$	$x = 0.500$	$x = 0.500$	$x = 0.500$	$x = 0.500$
		$y = 0.579$	$y = 0.591$	$y = 0.602$	$y = 0.612$	$y = 0.623$	$y = 0.633$	$y = 0.644$	$y = 0.654$	$y = 0.663$	$y = 0.673$
9 h	15 h	$x = 0.707$	$x = 0.707$	$x = 0.707$	$x = 0.707$	$x = 0.707$	$x = 0.707$	$x = 0.707$	$x = 0.707$	$x = 0.707$	$x = 0.707$
		$y = 0.473$	$y = 0.482$	$y = 0.491$	$y = 0.500$	$y = 0.509$	$y = 0.517$	$y = 0.525$	$y = 0.534$	$y = 0.542$	$y = 0.550$
8 h	16 h	$x = 0.866$	$x = 0.866$	$x = 0.866$	$x = 0.866$	$x = 0.866$	$x = 0.866$	$x = 0.866$	$x = 0.866$	$x = 0.866$	$x = 0.866$
		$y = 0.335$	$y = 0.341$	$y = 0.347$	$y = 0.354$	$y = 0.360$	$y = 0.366$	$y = 0.372$	$y = 0.377$	$y = 0.383$	$y = 0.389$
7 h	17 h	$x = 0.966$	$x = 0.966$	$x = 0.966$	$x = 0.966$	$x = 0.966$	$x = 0.966$	$x = 0.966$	$x = 0.966$	$x = 0.966$	$x = 0.966$
		$y = 0.173$	$y = 0.177$	$y = 0.180$	$y = 0.183$	$y = 0.186$	$y = 0.189$	$y = 0.192$	$y = 0.195$	$y = 0.198$	$y = 0.201$
6 h	18 h	$x = 1$	$x = 1$	$x = 1$	$x = 1$	$x = 1$	$x = 1$	$x = 1$	$x = 1$	$x = 1$	$x = 1$
		$y = 0$	$y = 0$	$y = 0$	$y = 0$	$y = 0$	$y = 0$	$y = 0$	$y = 0$	$y = 0$	$y = 0$
5 h	19 h	$x = 0.966$	$x = 0.966$	$x = 0.966$	$x = 0.966$	$x = 0.966$	$x = 0.966$	$x = 0.966$	$x = 0.966$	$x = 0.966$	$x = 0.966$
		$y = -0.173$	$y = -0.177$	$y = -0.180$	$y = -0.183$	$y = -0.186$	$y = -0.189$	$y = -0.192$	$y = -0.195$	$y = -0.198$	$y = -0.201$
4 h	20 h							$x = 0.866$	$x = 0.866$	$x = 0.866$	$x = 0.866$
								$y = -0.372$	$y = -0.377$	$y = -0.383$	$y = -0.389$

Appendix C

The equation of time and the Sun's declination

The equation of time

	January	February	March	April	May	June
1	−3 min 33 s	−13 min 35 s	−12 min 22 s	−3 min 54 s	+2 min 54 s	+2 min 12 s
2	−4 min 1 s	−13 min 42 s	−12 min 10 s	−3 min 37 s	+3 min 1 s	+2 min 2 s
3	−4 min 29 s	−13 min 49 s	−11 min 57 s	−3 min 19 s	+3 min 7 s	+1 min 52 s
4	−4 min 56 s	−13 min 55 s	−11 min 45 s	−3 min 1 s	+3 min 13 s	+1 min 42 s
5	−5 min 23 s	−14 min 0 s	−11 min 31 s	−2 min 44 s	+3 min 18 s	+1 min 32 s
6	−5 min 50 s	−14 min 5 s	−11 min 18 s	−2 min 27 s	+3 min 23 s	+1 min 21 s
7	−6 min 16 s	−14 min 8 s	−11 min 3 s	−2 min 10 s	+3 min 27 s	+1 min 10 s
8	−6 min 42 s	−14 min 11 s	−10 min 49 s	−1 min 53 s	+3 min 31 s	+0 min 59 s
9	−7 min 7 s	−14 min 13 s	−10 min 34 s	−1 min 37 s	+3 min 34 s	+0 min 47 s
10	−7 min 31 s	−14 min 14 s	−10 min 19 s	−1 min 20 s	+3 min 36 s	+0 min 35 s
11	−7 min 55 s	−14 min 14 s	−10 min 3 s	−1 min 4 s	+3 min 38 s	+0 min 23 s
12	−8 min 18 s	−14 min 14 s	−9 min 47 s	−0 min 49 s	+3 min 40 s	+0 min 11 s
13	−8 min 41 s	−14 min 12 s	−9 min 31 s	−0 min 33 s	+3 min 41 s	−0 min 1 s
14	−9 min 3 s	−14 min 10 s	−9 min 14 s	−0 min 18 s	+3 min 41 s	−0 min 14 s
15	−9 min 24 s	−14 min 8 s	−8 min 57 s	−0 min 3 s	+3 min 40 s	−0 min 26 s
16	−9 min 45 s	−14 min 4 s	−8 min 40 s	+0 min 11 s	+3 min 40 s	−0 min 39 s
17	−10 min 5 s	−14 min 0 s	−8 min 23 s	+0 min 25 s	+3 min 38 s	−0 min 52 s
18	−10 min 24 s	−13 min 55 s	−8 min 5 s	+0 min 39 s	+3 min 36 s	−1 min 5 s
19	−10 min 43 s	−13 min 50 s	−7 min 48 s	+0 min 52 s	+3 min 33 s	−1 min 18 s
20	−11 min 0 s	−13 min 44 s	−7 min 30 s	+1 min 5 s	+3 min 30 s	−1 min 31 s
21	−11 min 18 s	−13 min 37 s	−7 min 12 s	+1 min 17 s	+3 min 26 s	−1 min 44 s
22	−11 min 34 s	−13 min 30 s	−6 min 54 s	+1 min 29 s	+3 min 22 s	−1 min 58 s
23	−11 min 50 s	−13 min 22 s	−6 min 36 s	+1 min 40 s	+3 min 17 s	−2 min 11 s
24	−12 min 4 s	−13 min 13 s	−6 min 18 s	+1 min 51 s	+3 min 12 s	−2 min 24 s
25	−12 min 19 s	−13 min 4 s	−6 min 0 s	+2 min 2 s	+3 min 6 s	−2 min 37 s
26	−12 min 32 s	−12 min 54 s	−5 min 42 s	+2 min 12 s	+3 min 0 s	−2 min 49 s
27	−12 min 44 s	−12 min 44 s	−5 min 24 s	+2 min 21 s	+2 min 53 s	−3 min 2 s
28	−12 min 56 s	+12 min 33 s	−5 min 6 s	+2 min 30 s	+2 min 45 s	−3 min 14 s
29	−13 min 7 s	—	−4 min 48 s	+2 min 39 s	+2 min 38 s	−3 min 27 s
30	−13 min 17 s	—	−4 min 30 s	+2 min 46 s	+2 min 29 s	−3 min 39 s
31	−13 min 26 s	—	−4 min 12 s	—	+2 min 21 s	—

Opposite, below and on succeeding pages are two tables, the first for the equation of time and the second for the Sun's declination, both calculated for 12 h UT on each day of the year. Please note that these are mean values, valid for a few years around 2005. If greater accuracy is required, it will be necessary to refer to ephemerides. Moreover, the variations in the Earth's orbital elements mean that the equation of time and the Sun's declination also vary as the years go by, although these variations are negligible in the short term for sundial users. It would, however, not be advisable still to be using these tables in 2015!

	July	August	September	October	November	December
1	−3 min 50 s	−6 min 20 s	−0 min 2 s	+10 min 18 s	+16 min 24 s	+11 min 0 s
2	−4 min 2 s	−6 min 16 s	+0 min 17 s	+10 min 37 s	+16 min 26 s	+10 min 38 s
3	−4 min 13 s	−6 min 12 s	+0 min 37 s	+10 min 56 s	+16 min 26 s	+10 min 14 s
4	−4 min 24 s	−6 min 7 s	+0 min 56 s	+11 min 15 s	+16 min 26 s	+9 min 51 s
5	−4 min 34 s	−6 min 1 s	+1 min 16 s	+11 min 33 s	+16 min 24 s	+9 min 26 s
6	−4 min 45 s	−5 min 54 s	+1 min 37 s	+11 min 51 s	+16 min 22 s	+9 min 1 s
7	−4 min 54 s	−5 min 47 s	+1 min 57 s	+12 min 9 s	+16 min 20 s	+8 min 36 s
8	−5 min 4 s	−5 min 39 s	+2 min 18 s	+12 min 26 s	+16 min 16 s	+8 min 10 s
9	−5 min 13 s	−5 min 31 s	+2 min 39 s	+12 min 43 s	+16 min 11 s	+7 min 43 s
10	−5 min 21 s	−5 min 22 s	+3 min 0 s	+12 min 59 s	+16 min 6 s	+7 min 16 s
11	−5 min 29 s	−5 min 13 s	+3 min 21 s	+13 min 15 s	+16 min 0 s	+6 min 49 s
12	−5 min 37 s	−5 min 3 s	+3 min 42 s	+13 min 30 s	+15 min 53 s	+6 min 21 s
13	−5 min 44 s	−4 min 52 s	+4 min 3 s	+13 min 45 s	+15 min 45 s	+5 min 53 s
14	−5 min 51 s	−4 min 41 s	+4 min 25 s	+13 min 59 s	+15 min 36 s	+5 min 24 s
15	−5 min 57 s	−4 min 29 s	+4 min 46 s	+14 min 12 s	+15 min 26 s	+4 min 56 s
16	−6 min 3 s	−4 min 17 s	+5 min 7 s	+14 min 25 s	+15 min 15 s	+4 min 26 s
17	−6 min 8 s	−4 min 5 s	+5 min 29 s	+14 min 38 s	+15 min 4 s	+3 min 57 s
18	−6 min 13 s	−3 min 52 s	+5 min 50 s	+14 min 50 s	+14 min 52 s	+3 min 28 s
19	−6 min 17 s	−3 min 38 s	+6 min 11 s	+15 min 1 s	+14 min 38 s	+2 min 58 s
20	−6 min 21 s	−3 min 24 s	+6 min 33 s	+15 min 11 s	+14 min 24 s	+2 min 28 s
21	−6 min 24 s	−3 min 9 s	+6 min 54 s	+15 min 21 s	+14 min 10 s	+1 min 58 s
22	−6 min 27 s	−2 min 54 s	+7 min 15 s	+15 min 31 s	+13 min 54 s	+1 min 28 s
23	−6 min 29 s	−2 min 39 s	+7 min 36 s	+15 min 39 s	+13 min 38 s	+0 min 58 s
24	−6 min 30 s	−2 min 23 s	+7 min 57 s	+15 min 47 s	+13 min 20 s	+0 min 28 s
25	−6 min 31 s	−2 min 7 s	+8 min 18 s	+15 min 54 s	+13 min 3 s	−0 min 1 s
26	−6 min 32 s	−1 min 50 s	+8 min 38 s	+16 min 1 s	+12 min 44 s	−0 min 31 s
27	−6 min 31 s	−1 min 33 s	+8 min 59 s	+16 min 7 s	+12 min 25 s	−1 min 1 s
28	−6 min 30 s	−1 min 16 s	+9 min 19 s	+16 min 12 s	+12 min 5 s	−1 min 30 s
29	−6 min 29 s	−0 min 58 s	+9 min 39 s	+16 min 16 s	+11 min 44 s	−2 min 0 s
30	−6 min 27 s	−0 min 40 s	+9 min 59 s	+16 min 19 s	+11 min 22 s	−2 min 29 s
31	−6 min 24 s	−0 min 21 s	—	+16 min 22 s	—	−2 min 57 s

Declination of Sun

	January	February	March	April	May	June
1	−22° 58′	−16° 59′	−7° 26′	+4° 41′	+15° 12′	+22° 06′
2	−22° 53′	−16° 42′	−7° 04′	+5° 04′	+15° 29′	+22° 14′
3	−22° 47′	−16° 24′	−6° 41′	+5° 27′	+15° 47′	+22° 22′
4	−22° 41′	−16° 07′	−6° 18′	+5° 50′	+16° 05′	+22° 29′
5	−22° 34′	−15° 48′	−5° 54′	+6° 13′	+16° 22′	+22° 35′
6	−22° 27′	−15° 30′	−5° 31′	+6° 36′	+16° 39′	+22° 42′
7	−22° 19′	−15° 11′	−5° 08′	+6° 58′	+16° 55′	+22° 47′
8	−22° 11′	−14° 52′	−4° 44′	+7° 21′	+17° 12′	+22° 53′
9	−22° 03′	−14° 33′	−4° 21′	+7° 43′	+17° 28′	+22° 58′
10	−21° 54′	−14° 13′	−3° 57′	+8° 05′	+17° 43′	+23° 02′
11	−21° 45′	−13° 54′	−3° 34′	+8° 27′	+17° 59′	+23° 07′
12	−21° 35′	−13° 34′	−3° 10′	+8° 49′	+18° 14′	+23° 11′
13	−21° 25′	−13° 14′	−2° 47′	+9° 11′	+18° 29′	+23° 14′
14	−21° 14′	−12° 53′	−2° 23′	+9° 33′	+18° 43′	+23° 17′
15	−21° 03′	−12° 33′	−1° 59′	+9° 54′	+18° 57′	+23° 20′
16	−21° 52′	−12° 12′	−1° 35′	+10° 16′	+19° 11′	+23° 22′
17	−20° 40′	−11° 51′	−1° 12′	+10° 37′	+19° 25′	+23° 23′
18	−20° 28′	−11° 30′	−0° 48′	+10° 58′	+19° 36′	+23° 25′
19	−20° 15′	−11° 08′	−0° 24′	+11° 18′	+19° 51′	+23° 26′
20	−20° 02′	−10° 47′	−0° 00′	+11° 39′	+20° 04′	+23° 26′
21	−19° 49′	−10° 25′	+0° 23′	+11° 59′	+20° 16′	+23° 26′
22	−19° 35′	−10° 03′	+0° 47′	+12° 20′	+20° 28′	+23° 26′
23	−19° 21′	−9° 41′	+1° 10′	+12° 40′	+20° 39′	+23° 25′
24	−19° 07′	−9° 19′	+1° 34′	+12° 59′	+20° 50′	+23° 24′
25	−18° 52′	−8° 57′	+1° 58′	+13° 19′	+21° 01′	+23° 23′
26	−18° 37′	−8° 34′	+2° 21′	+13° 38′	+21° 12′	+23° 21′
27	−18° 21′	−8° 12′	+2° 45′	+13° 57′	+21° 22′	+23° 18′
28	−18° 06′	−7° 49′	+3° 08′	+14° 16′	+21° 31′	+23° 16′
29	−17° 50′	—	+3° 32′	+14° 35′	+21° 41′	+23° 13′
30	−17° 33′	—	+3° 55′	+14° 53′	+21° 50′	+23° 09′
31	−17° 16′	—	+4° 18′	—	+21° 58′	—

	July	August	September	October	November	December
1	+23° 05′	+17° 55′	+8° 09′	−3° 20′	−14° 33′	−21° 52′
2	+23° 01′	+17° 40′	+7° 47′	−3° 43′	−14° 52′	−22° 01′
3	+22° 56′	+17° 24′	+7° 25′	−4° 06′	−15° 11′	−22° 09′
4	+22° 51′	+17° 08′	+7° 03′	−4° 30′	−15° 29′	−22° 17′
5	+22° 45′	+16° 52′	+6° 40′	−4° 53′	−15° 48′	−22° 25′
6	+22° 39′	+16° 36′	+6° 18′	−5° 16′	−16° 06′	−22° 32′
7	+22° 33′	+16° 19′	+5° 55′	−5° 39′	−16° 23′	−22° 39′
8	+22° 26′	+16° 02′	+5° 33′	−6° 02′	−16° 41′	−22° 45′
9	+22° 19′	+15° 45′	+5° 10′	−6° 25′	−16° 58′	−22° 51′
10	+22° 11′	+15° 27′	+4° 48′	−6° 47′	−17° 15′	−22° 57′
11	+22° 03′	+15° 09′	+4° 25′	−7° 10′	−17° 32′	−23° 02′
12	+21° 55′	+14° 51′	+4° 02′	−7° 32′	−17° 48′	−23° 06′
13	+21° 46′	+14° 33′	+3° 39′	−7° 55′	−18° 04′	−23° 10′
14	+21° 37′	+14° 15′	+3° 16′	−8° 17′	−18° 19′	−23° 14′
15	+21° 28′	+13° 56′	+2° 53′	−8° 39′	−18° 35′	−23° 17′
16	+21° 18′	+13° 37′	+2° 30′	−9° 01′	−18° 50′	−23° 20′
17	+21° 08′	+13° 18′	+2° 07′	−9° 23′	−19° 05′	−23° 22′
18	+20° 57′	+12° 58′	+1° 43′	−9° 45′	−19° 19′	−23° 24′
19	+20° 47′	+12° 39′	+1° 20′	−10° 07′	−19° 33′	−23° 25′
20	+20° 35′	+12° 19′	+0° 57′	−10° 28′	−19° 46′	−23° 26′
21	+20° 24′	+11° 59′	+0° 33′	−10° 50′	−20° 00′	−23° 26′
22	+20° 12′	+11° 39′	+0° 10′	−11° 11′	−20° 13′	−23° 26′
23	+20° 00′	+11° 19′	−0° 13′	−11° 32′	−20° 25′	−23° 26′
24	+19° 47′	+10° 58′	−0° 37′	−11° 53′	−20° 37′	−23° 25′
25	+19° 34′	+10° 38′	−1° 00′	−12° 14′	−20° 49′	−23° 23′
26	+19° 21′	+10° 17′	−1° 23′	−12° 34′	−21° 01′	−23° 21′
27	+19° 08′	+9° 56′	−1° 47′	−12° 55′	−21° 12′	−23° 19′
28	+18° 54′	+9° 35′	−2° 10′	−13° 15′	−21° 22′	−23° 16′
29	+18° 40′	+9° 13′	−2° 33′	−13° 35′	−21° 32′	−23° 12′
30	+18° 25′	+8° 52′	−2° 57′	−13° 54′	−21° 42′	−23° 09′
31	+18° 10′	+8° 30′	—	−14° 14′	—	−23° 04′

Formulae for diurnal motion

We can directly establish formulae for diurnal motion through the use of spherical trigonometry. If we confine ourselves (as below) to plane trigonometry, we must implicitly reconstitute the formulae for spherical triangles to achieve the same end.

D.1 The semi-diurnal arc

Figure D.1 represents the Celestial Sphere, centered on O, as seen looking eastwards (i.e. north is to the left; south to the right). Each day, the Sun describes a circle centered on O' (M' = midnight, L = sunrise, M = midday/noon, C = sunset).

That part of the circle (LMC) above the horizon is called the diurnal arc. Points D and O are on the north–south meridian NS; the position of D varies according to the declination δ:

$$\delta = \widehat{AM} = \widehat{AOM}$$

The planes of the diurnal arc and the Equator make an angle equal to $90° - \varphi$ (φ = latitude) with that of the horizon. The semi-diurnal arc:

$$H_0 = \widehat{LM} = \widehat{LO'M} = \widehat{MO'C} = \widehat{MC}$$

In the triangle $O'DC$ (Figure D.2):

$$\widehat{DO'C} = 180° - H_0$$

$$O'D = O'C \cos \widehat{DO'C} = -O'C \cos H_0$$

In the triangle $MO'O$ (Figure D.3):

$$OM = OA = \text{radius of the sphere}$$

$$O'M = OM \cos \delta = OA \cos \delta$$

but $O'C = O'M = OA \cos \delta$, whence:

$$\cos H_0 = \frac{-O'D}{O'C} = \frac{-O'D}{O'M} = \frac{-O'D}{OA \cos \delta}$$

In the triangle $DO'O$ (Figure D.3):

$$O'O = O'D \tan \widehat{O'DO} = O'D \tan(90° - \varphi) = O'D/\tan \varphi$$

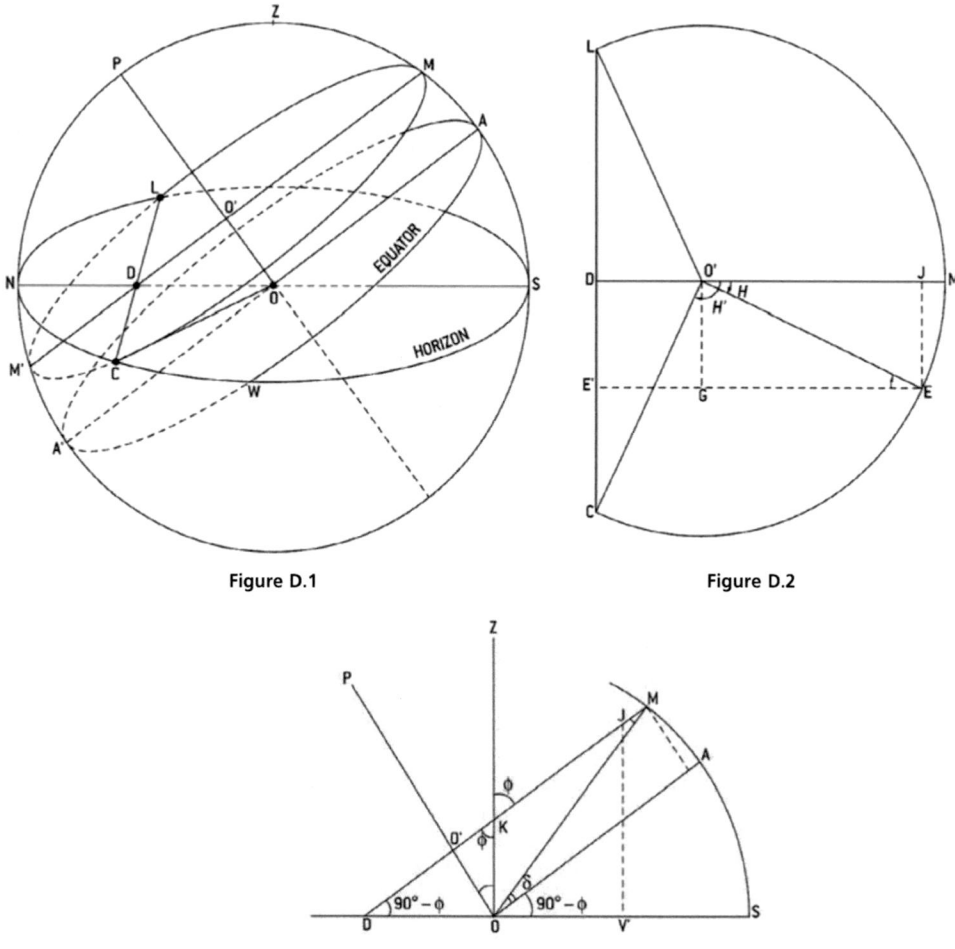

Figure D.1 Figure D.2

Figure D.3

In the triangle MO′O (Figure D.3):

$$OO' = OM \sin \delta = OA \sin \delta$$

Comparing the two expressions of $OO' = O'O$, we have:

$$O'D/\tan \varphi = OA \sin \delta$$

whence $O'D = OA \tan \varphi \sin \delta$.

Finally:

$$\cos H_0 = \frac{-O'D}{OA \cos \delta} = \frac{-OA \tan \varphi \sin \delta}{OA \cos \delta} = -\tan \varphi \tan \delta$$

$$\cos H_0 = -\tan \varphi \tan \delta \qquad\qquad (D.1)$$

D.2   The azimuth of a celestial body at rising and setting

The two azimuths of the Sun at its rising and setting are equal and of opposite sign, assuming the same declination δ. They are reckoned clockwise in the horizontal plane from point S (south) (Figure D.1).

$$A_0 \text{ at setting} = \widehat{SC} = \widehat{SOC}$$

In the triangle DOC:

$$OC = OA = \text{radius of the sphere and } \widehat{DOC} = 180° - A_0$$

$$OD = OC \cos(180° - A_0) = -OA \cos A_0$$

$$\text{and} \quad \cos A_0 = \frac{-OD}{OA}$$

In the triangle DO'O:

$$OO' = OD \sin(90° - \varphi) = OD \cos \varphi \text{ and } OD = \frac{OO'}{\cos \varphi}$$

But $OO' = OA \sin \delta$, whence:

$$OD = \frac{OA \sin \delta}{\cos \varphi} \quad \text{and} \quad \frac{OD}{OA} = \frac{\sin \delta}{\cos \varphi}$$

Finally:

$$\cos A_0 = \frac{-\sin \delta}{\cos \varphi} \qquad\qquad (D.2)$$

D.3   The altitude of the Sun

To calculate the altitude of the Sun, we need to know its hour angle H. Let us suppose that the Sun is at E on the diurnal arc (Figure D.2). We drop perpendiculars EE', O'G and EJ. Point J, at the same distance as point E from CDL, is at the same linear altitude (JV') as E (the verticals EV and JV' are not shown in Figure D.1: see Figure D.4). We need only calculate JV' to find EV.

We see that (Figure D.3):

$$JV' = DJ \sin(90° - \varphi) = DJ \cos \varphi$$

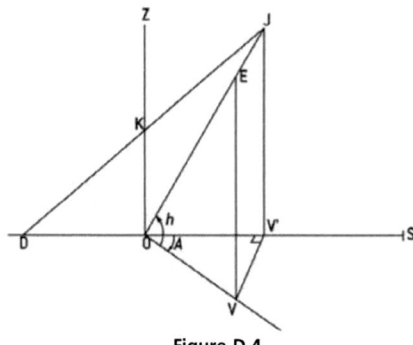

Figure D.4

but $DJ = E'E = E'G + GE$

$$E'G = O'D = OA \tan \varphi \sin \delta$$

$$GE = O'J = O'E \cos H = O'M \cos H$$

We have already found that $O'M = OA \cos \delta$, whence:

$$GE = OA \cos \delta \cos H$$

and

$$EV = JV' = (OA \tan \varphi \sin \delta + OA \cos \delta \cos H) \cos \varphi$$

$$EV = OA(\sin \varphi \sin \delta + \cos \varphi \cos \delta \cos H)$$

The angular altitude h of the Sun at E (Figure D.4) is $\widehat{EOV}$; now,

$$EV = OE \sin h \quad \text{or} \quad \sin h = \frac{EV}{OE}$$

but $OE = OM = OA = $ radius of the sphere, whence $\sin h = \dfrac{EV}{OA}$.
Finally:

$$\sin h = \sin \delta \sin \varphi + \cos \varphi \cos \delta \cos H \qquad \text{(D.3)}$$

D.4 The azimuth of the Sun

Two cases are possible:

- We already know the altitude (Figure D.4). As we have already seen (Figure D.2):

$$VV' = EJ = O'E \sin H = OA \cos \delta \sin H$$

In the triangle OVV' (Figure D.4): $VV' = OV \sin A$
but $OV = OE \cos h = OA \cos h$

whence $VV' = OA \cos h \sin A$
Comparing the two expressions of VV':

$$VV' = OA \cos \delta \sin H = OA \cos h \sin A$$

or $\cos \delta \sin H = \cos h \sin A$, and finally:

$$\sin A = \frac{\cos \delta \sin H}{\cos h} \tag{D.4}$$

- We know H, but we do not know the value for h.
  In Figure D.4, in the triangle OVV': $OV'/VV' = \cotan A$.
  We already know that $VV' = OA \cos \delta \sin H$
  Note that OV' is the projection of KJ (Figure D.3) onto the horizontal plane:

  $$OV' = KJ \cos(90° - \varphi) = KJ \sin \varphi$$

But $KJ = O'J - O'K$

$$O'J = O'E \cos H = OA \cos \delta \cos H$$

$$O'K = OO' \tan(90° - \varphi) = OA \sin \delta / \tan \varphi$$

whence:

$$OV' = OA(\cos \delta \cos H - \sin \delta / \tan \varphi) \sin \varphi$$

Finally:

$$\tan A = \frac{OA \cos \delta \sin H}{OA(\cos \delta \cos H - \sin \delta \cotan \varphi) \sin \varphi}$$

or, dividing by $\cos \delta$:

$$\tan A = \frac{\sin H}{\sin \varphi \cos H - \cos \varphi \tan \delta} \tag{D.5}$$

D.5   Prime vertical passage

The prime vertical of a place is the vertical plane passing through east and west, azimuth $-90°$. From Eqn. (D.5), we can deduce the Sun's hour angle as it crosses this plane, noting that:

$$\frac{\cos A}{\sin A} = \frac{\sin \varphi \cos H - \cos \varphi \tan \delta}{\sin H}$$

With $A = +90°$, we deduce that:

$$\sin \varphi \cos H = \cos \varphi \tan \delta$$

whence:

$$\cos H = \frac{\tan \delta}{\tan \varphi} \tag{D.6}$$

Demonstrations

Certain formulae used in this book, for example to draw the faces of sundials, may be demonstrated using only trigonometry and spatial geometry. Here are some examples.

5.1 Meridional sundials showing mean solar time

On certain sundials we see a curve, in the form of a figure of eight, astride the noon line. This is the mean noon curve. When the shadow of the style or the spot of light cast by an eyelet falls upon the figure-of-eight, we can directly read local mean time (or Universal Time if the creator of the dial has taken into account the equation of time and the local longitude). Some sundial makers have even calculated a curve for each hour of daylight. Although the calculation of such a curve is beyond the scope of this book, investigating it will give an insight into the nature of the equation of time, which is the difference between *true* solar time and *mean* solar time. Since a sundial in fact measures the Sun's hour angle, we can also say that the equation of time represents the difference between the hour angle of the *true* Sun and that of the *mean* Sun.

To an astronomer, the term 'hour angle' has a very precise definition: it is the difference between local sidereal time and the right ascension of the celestial object in question (here, the Sun). Sidereal time itself is the hour angle of the vernal equinox (the First Point of Aries). As the Earth rotates, this point makes a complete circuit of the sky in 23 h 56 min 04 s. In some observatories there is a clock showing sidereal time, often mounted beside one showing Universal Time: the two clocks are used to locate celestial objects of known right ascension by referring to the time.

Let us call the hour angles of the mean Sun and of the true Sun H_m and H_t respectively. By definition, the equation of time E is equal to:

$$E = H_t - H_m$$

Now, if T is the sidereal time,

$$E = (T - \alpha_t) - (T - \alpha_m)$$

where α_m and α_t are respectively the right ascension of the mean Sun and the right ascension of the true Sun. We can therefore write:

$$E = \alpha_m - \alpha_t$$

Now we have a new definition of the equation of time: it is the difference between the right ascension of the *mean* Sun and that of the *true* Sun. This will perhaps give a better insight into the definition of the mean Sun: it is an imaginary Sun moving along the Celestial Equator (its declination therefore being zero), and returning to the local meridian after exactly 24 h 00 min 00 s. It completes its course through the 360° of the Celestial Equator in one year (365.2422 days).

As for the true Sun, it oscillates from one side of the mean Sun to the other during the year (which is what the figure-of-eight curve represents), because its right ascension does not change in a uniform way. In fact:

$$\tan \alpha_t = \cos \varepsilon \tan \lambda$$

ε being the obliquity of the ecliptic. This formula alone practically sums up the equation of time! The term $\cos \varepsilon$ represents what was termed in Chapter 1.7 the *reduction to the Equator*, i.e. the projection of the true Sun (on the ecliptic) onto the Equator. The term $\tan \lambda$ represents the *equation of the centre* (also in Chapter 1.7), since (implicitly) the Sun's apparent longitude λ shows the inequality due to Kepler's laws.

E.2 Formula for the direct south sundial based on the horizontal sundial

In Figure E.1, H'' is the angle on the vertical dial between an hour line and the noon line. Point A is also the extremity of the noon line of the horizontal dial, on which the angle between an hour line and noon is calculated by:

$$\tan H' = \sin \varphi \tan H$$

On the vertical dial, $AB/AD = \tan H''$, and $AD = DC \sin \varphi$. Also,

$$AB/AC = \tan H' = \sin \varphi \tan H$$

Therefore

$$AB = AC \sin \varphi \tan H$$

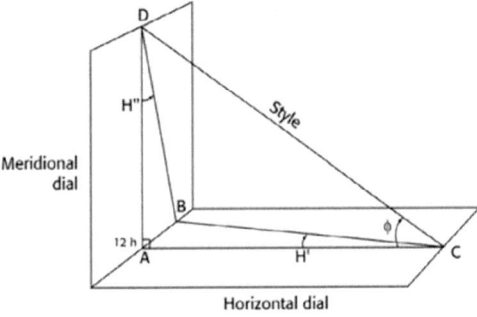

Figure E.1

Now,
$$AC = DC \cos \varphi$$
so
$$AB = DC \cos \varphi \sin \varphi \tan H$$
Finally:
$$\tan H'' = DC \cos \varphi \sin \varphi \tan H / DC \sin \varphi$$
or, simplifying:
$$\tan H'' = \cos \varphi \tan H$$

E.3   Drawing an analemmatic sundial using an equatorial sundial

Take a functioning circular equatorial sundial of radius $PE = a$ (Figure E.2). Point O is the point on the ground vertically beneath point P (the centre of the equatorial dial). With O at the centre, we draw on the ground a circle of the same radius and an ellipse: OA is the semi-major axis (Figure E.3). Since the plane of the equatorial sundial is parallel to the Celestial Equator, $OB = PB \sin \varphi$. OB is the semi-minor axis (marked b) of the sundial on the ground. $b = a \sin \varphi$. On the circle, point I makes an angle H with the meridian, reckoned positively from north, clockwise: H is the hour angle of the Sun. The x abscissa of point I reckoned along the semi-major axis is equal to OI sin H. Now, $OI = OA = PE = a$. Whence:
$$x = a \sin H$$
If Y is the ordinate of point I, then the ordinate y on the ellipse is $y = Y \times b/a$. Now $Y = a \cos H$, whence $y = b \cos H$. Since $b = a \sin \varphi$,
$$y = a \sin \varphi \cos H$$
Figure E.4 shows the equatorial sundial seen side-on, PK being the 'useful' part of the polar style: the image of K should always fall on B, whatever the date. There-

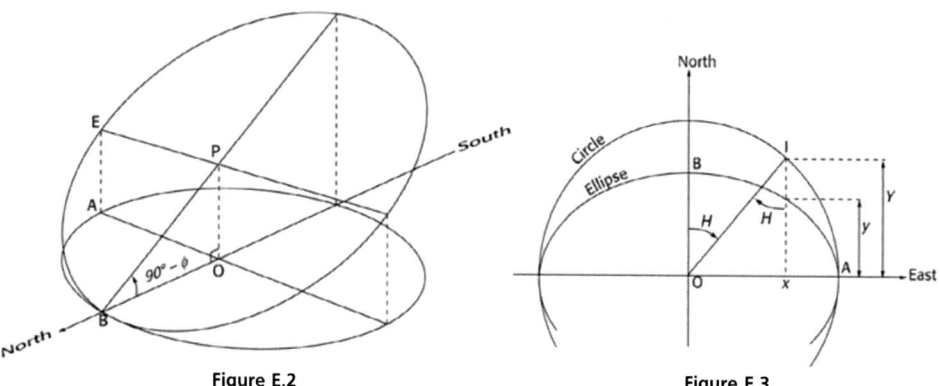

Figure E.2   Figure E.3

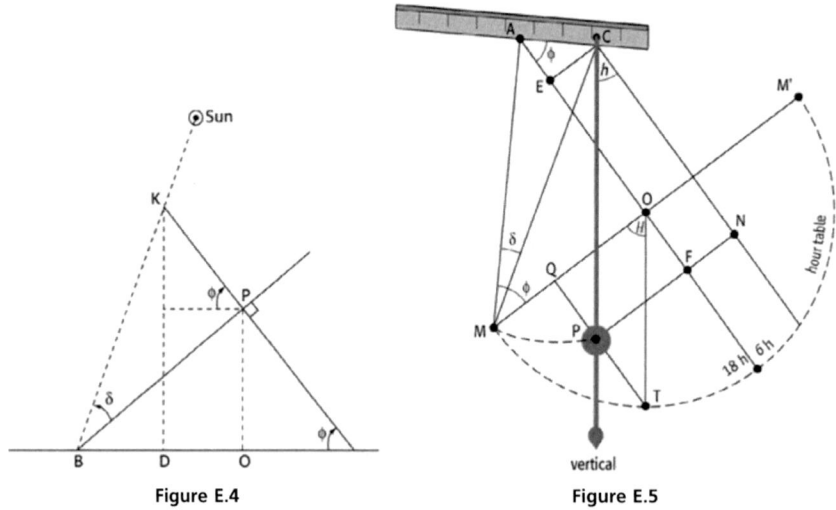

Figure E.4   Figure E.5

fore K should be movable along the polar axis. Angle PBK is therefore equal to the Sun's declination δ. The displacement of the gnomon on the north-south axis is here equal to OD. PK = $a \tan \delta$. Also, OD = PK cos φ, whence, finally:

$$OD = a \cos \varphi \tan \delta$$

E.4 Demonstration of the Saint Rigaud Capuchin sundial

In the triangle OQT (Figure E.5),

$$\cos H = \frac{OQ}{OT}$$

Now,

$$OQ = PN - FN$$

$$PN = CP \sin h = CM \sin h$$

$$CM \cos \delta = MA = \frac{OM}{\cos \varphi}$$

whence

$$CM = \frac{OM}{\cos \varphi \cos \delta}$$

Therefore

$$FN = EC = AC \sin \varphi$$

and

$$AC = AM \tan \delta = \frac{OM \tan \delta}{\cos \varphi}$$

$$\cos H = \frac{OQ}{OM} = \frac{\dfrac{OM \sin h}{\cos \varphi \cos \delta} - \dfrac{OM \sin \varphi \tan \delta}{\cos \varphi}}{OM}$$

$$\cos H = \frac{\sin h - \sin \varphi \sin \delta}{\cos \varphi \cos \delta}$$

or

$$\sin h = \sin \varphi \sin \delta + \cos \varphi \cos \delta \cos H \qquad (E.1)$$

Equation (E.1) gives the altitude of the Sun as a function of time (H), latitude (φ) and date (δ). All altitude sundials are based on this principle: they may be seen as abacuses which graphically represent the resolution of the classic formula (Equation E.1).

Small problems in gnomonics

F.1 What is the relationship between the hour angle and the azimuth?

At what time (i.e. local clock time) does the Sun reach a certain azimuth (A) for a given location and day?

The Sun's azimuth A, as a function of latitude (φ), hour angle (H) and declination (δ), is obtained (see Appendix D, page 153 for table of formulae for diurnal motion) by:

$$\tan A = \frac{\sin H}{\sin \varphi \cos H - \cos \varphi \tan \delta} \qquad \text{(F.1)}$$

which may be written:

$$C_1 \cos H - C_2 \sin H = C_3$$

where $C_1 = \sin \varphi$; $C_2 = \cotan A$; $C_3 = \cos \varphi \tan \delta$. To solve the equation, we use an auxiliary angle M such that:

$$\tan M = \sin \varphi \tan A$$

(M must have the same sign as A; if not, add $\pm 180°$)

$$\sin(M - H) = \frac{\cos \varphi \tan \delta \sin M}{\sin \varphi}$$

Example 1: $A = 30°$, $\delta = +23°.44$, $\varphi = 48°$: we obtain $M = 23°.222$, and $H = 14°.367$.

Example 2: $A = 105°$, $\delta = +23°.44$, $\varphi = 48°$: we obtain $M = 109°.827$, and $H = 88°.282$.

Example 3: $A = -60°$, $\delta = +23°.44$, $\varphi = 48°$: we obtain $M = -52°.156$, and $H = -34°.201$.

F.2 Are azimuth and hour angle equal?

We sometimes read that all we have to do to make a sundial is to plant a stick in the ground and draw straight lines around it, $15°$ apart: when the shadow of the stick lies along one of these lines, we can read the time ... Such a sundial will work only at the poles!

If we draw an array of straight lines 15° apart around our stick (assuming we have determined the local meridian), we are measuring the angle between the direction of the Sun and due south, the Sun's azimuth:

$$\tan A = \frac{\sin H}{\sin \varphi \cos H - \cos \varphi \tan \delta} \qquad (F.2)$$

where H is the Sun's hour angle, φ the local latitude, and δ the Sun's declination. For latitude 48°, at the summer solstice ($\delta = 23°.44$) in the afternoon, we calculate the values in Table F.1.

Table F.1		
$H = 0°$ [12 h]	$A = 0°$	$A - H = 0°$
$H = 15°$ [13 h]	$A = 31°.179$	$A - H = 16°.179$
$H = 30°$ [14 h]	$A = 54°.742$	$A - H = 24°.742$
$H = 45°$ [15 h]	$A = 71°.589$	$A - H = 26°.589$
$H = 60°$ [16 h]	$A = 84°.627$	$A - H = 24°.627$
$H = 75°$ [17 h]	$A = 95°.780$	$A - H = 20°.780$
$H = 90°$ [18 h]	$A = 106°.178$	$A - H = 16°.178$
$H = 105°$ [19 h]	$A = 116°.541$	$A - H = 11°.541$

These results show that, if we assume the azimuth to be equal to the hour angle, we can be in error by more than 26°: more than 1 h 45 min! The error is due to the fact that the azimuth and the hour angle are two different coordinates, measured in different planes; the azimuth in the plane of the horizon, and the hour angle in that of the Celestial Equator.

If we are at the North Pole, for example, with $\varphi = 90°$, the formula above becomes $\tan A = \tan H$, i.e. $A = H$. Easily explained: at the North Pole, the Celestial Equator coincides with the horizon.

F.3   What is the length of the shadow of a gnomon in the course of a day?

Should we need to calculate the length of the shadow of a gnomon at a given time of day on a given date—for example, of a vertical 10-meter mast at 16 h (clock time) on June 15 at Toulon—we first find the altitude of the Sun using $\sin h = \sin \varphi \sin \delta + \cos \varphi \cos \delta \cos H$ (see Equation D.3 in Appendix D, page 153) where φ is the local latitude, and δ the Sun's declination as in the tables in Appendix C, page 148. The hour angle is the difference between Universal Time and the time of the passage of the Sun across the local meridian. We must first of all convert the time to Universal Time: as it is June 15, Summer Time is in force, and we need to subtract 2 hours to arrive at 14 h UT. The latitude of Toulon is 43° 07′ and the longitude −23 min 44 s. On June 15, the value for the equation of time is −0 min 26 s. Therefore, the Sun crosses the meridian at Toulon on that day at: 12 h −(−0 min 26 s) − 23 min 44 s = 11 h 36 min 42 s UT.

So the hour angle is 14 h UT − 11 h 36 min 42 s UT = 2 h 23 min 18 s: in degrees, +35° 49′ 30″. The altitude h of the Sun is therefore 54° 30′ 20″. The length of the shadow of the 10-meter mast is therefore equal to $(10/\tan h) = 7.1$ m. If we had to calculate the length of the shadow at 10 h UT, the hour angle would be negative ($H = -24°\ 10′\ 30″$).

When working with these kinds of problems, we must always be aware of the signs of the different quantities (Sun's declination, equation of time, longitude, hour angle).

F.4   For how long will a vertical direct south sundial be illuminated?

For a wall facing due south (i.e. aligned east–west) to be sunlit, the Sun must be in front of the wall and above the horizon. To calculate the length of time during which the south-facing side is sunlit, we must first determine when the Sun crosses the prime vertical (the vertical plane passing through east and west), first to the east and then to the west. The hour angle of the Sun at the time of prime vertical passage is equal to

$$\cos H = \frac{\tan \delta}{\tan \varphi}$$

The total duration in hours of the illumination is therefore $(2H/15)$.

For example, at the summer solstice ($\delta = +23°.44$), at Nice (latitude 43° 42′), the duration of illumination of the south-facing side is 8 h 24 min. At Lille (latitude 50° 39′), the value is 9 h 13 min. Conclusion: the south-facing side of a wall in Lille is lit for longer in summer than the south-facing side of a wall in Nice.

Returning to the wall in Nice: is it lit for longer at the summer solstice or at the winter solstice? The Sun does not cross the prime vertical in winter since it rises in the south-east (and sets in the south-west). The wall is therefore lit from sunrise to sunset, i.e. $(2H_0/15)$, a total of 8 h 44 min. Conclusion: the south-facing wall in Nice is lit for longer at the winter solstice than it is at the summer solstice! Is this also the case in Lille?

F.5   On what dates is a vertical direct north sundial sunlit, and for how long?

First we must understand that, when the Sun ceases to illuminate the north-facing side, it will illuminate the south-facing side, and *vice versa*. For the north-facing side to be lit in France, the Sun has to rise in the north-east and set in the north-west, which is possible only if its declination is positive, i.e. from the spring equinox to the autumn equinox. We can therefore deduce that a surface facing due north can be lit during a six-month period approximately. But for how long during the day? Let us take as an example a latitude of $\varphi = 48°$ at the summer solstice ($\delta = +23°.44$). When the Sun rises in the north-east, its hour angle H_0 at the time of rising is equal to:

$$\cos H_0 = -\tan \varphi \tan \delta$$

which, here, equals $-118°.79$. The north-facing side is lit from the moment of sunrise until the Sun reaches azimuth $90°$, which is when it is due east (prime vertical passage). We know that, at this time,

$$\cos H = \frac{\tan \delta}{\tan \varphi}$$

In the present case, we obtain $H = -67°.02$. From this we readily conclude that the north-facing side will be lit from sunrise until the Sun is due east, i.e. for $(H_0 - H) = -51°.77 = 3\,\mathrm{h}\,27\,\mathrm{min}$, on the day of the summer solstice.

   If we repeat the calculation for the afternoon, we also arrive at a value of $3\,\mathrm{h}$ $27\,\mathrm{min}$, since the north-facing side will be lit from the time when the azimuth of the Sun is $+90°$ until sunset. So, on the day of the summer solstice, the north-facing side is lit for a total of $6\,\mathrm{h}\,54\,\mathrm{min}$: a far from negligible amount of time!

F.6   Under what circumstances can we observe the retrograde motion of the shadow of a gnomon?

At certain times of the year, we can perform a very interesting experiment involving the retrograde motion of the shadow of a gnomon, i.e. observing that the direction of the rotation of the shadow reverses itself. This phenomenon is linked to a particular feature of the Sun's azimuth in tropical zones: when the value for the declination of the Sun is greater than that of the local latitude, the Sun's azimuth has the same value twice in the morning and twice in the afternoon. Just one condition has to be satisfied: the value for the declination of the Sun must be greater (in absolute terms) than that of the local latitude, and of the same sign. Let us take a specific example, for a location at latitude $20°$. First, we determine the local meridian and draw the curve described by the tip of the shadow of a gnomon on 21 June ($\delta = +23°.44$). We set up a system of axes based on the gnomon: the y coordinates are reckoned positively northwards, and the x coordinates are reckoned positively eastwards. Taking the length of the gnomon to be $50\,\mathrm{cm}$, and using the formula $y = f(x)$

$$y = \frac{-a \sin \varphi \cos \varphi + \sin \delta \sqrt{[x^2(\cos^2 \varphi - \sin^2 \delta) + a^2 \cos^2 \delta]}}{(\sin^2 \delta - \cos^2 \varphi)}$$

we obtain:

$$x = 0 \text{ cm} \quad \text{and} \quad y = -3 \text{ cm}$$
$$x = 10 \text{ cm} \quad \text{and} \quad y = -3.44 \text{ cm}$$
$$x = 30 \text{ cm} \quad \text{and} \quad y = -6.64 \text{ cm}$$
$$x = 50 \text{ cm} \quad \text{and} \quad y = -12.18 \text{ cm}$$
$$x = 70 \text{ cm} \quad \text{and} \quad y = -19.10 \text{ cm}$$

We must be very precise with locating the coordinates. We verify carefully that the shadow at solar noon ($x = 0$) points south. On the day of the summer solstice (21 June), in a patient frame of mind since the phenomenon is slow and not easy to follow, we observe the motion of the shadow of the gnomon. Since it moves in the opposite direction to the Sun's motion, the shadow of the gnomon will move anti-clockwise after meridian passage. At first, we can see that the shadow elongates (which it never ceases to do); then, its rotational motion seems to slow, becomes stationary, and finally changes direction and retraces the same path described earlier. The best method is to mark the shadow with a line every 30 minutes: this will indeed show that the direction of the shadow has gone into reverse.

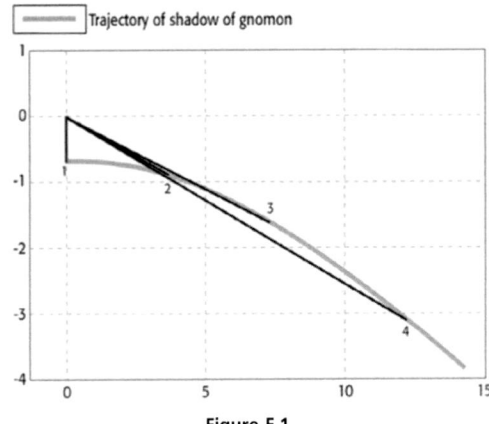

Figure F.1

Practicalities of constructing a sundial

G.1 Conversion of degrees, minutes and seconds into hours, minutes and seconds

The longitude of the Paris Observatory is $2°\ 20'\ 15''$ E. First, we convert the longitude into decimal form, dividing the minutes by 60 and the seconds by 3600:

$$2 + (20/60) + (15/3600) = 2°.3375$$

To obtain the longitude in hours, we divide by $15°$ (since $15° = 1\,h$):

$$2°.3375 \longrightarrow 0.1558333 \text{ decimal hours}$$

To convert this into hours, minutes and seconds, we multiply the result by 60:

$$0.1558333 \text{ decimal hours} \longrightarrow 9.35 \text{ minutes}$$

To obtain seconds, we subtract the minutes and again multiply the result by 60:

$$0.35 \text{ minutes} \longrightarrow 21 \text{ seconds}$$

So the longitude of the Paris Observatory expressed in terms of time is:

$$-0\,h\,9\,min\,21\,s \text{ (note the sign!)}$$

To express the longitude in time in degrees, minutes and seconds, the principle remains the same: we divide the minutes by 60 and the seconds by 3600 to obtain the longitude in decimal hours:

$$0\,h\,9\,min\,21\,s = 0\,h + (9/60) + (21/3600) = 0.1558333 \text{ hours}$$

Multiplying by 15, we obtain $2°.3375$. Subtracting the degrees and multiplying by 60, we obtain the minutes, i.e. $20'$. Subtracting the minutes and multiplying by 60, we obtain the seconds, i.e. $15''$.

We must be careful not to mix up the minutes and the seconds of time with angular minutes and seconds! For example, it is incorrect to write that a certain sporting record was broken in $12'\ 45''$: this must be written as $12\,min\,45\,s$.

G.2 Calculating the Sun's local meridian passage with astronomical ephemerides and the Internet

Most astronomical ephemerides give times (UT) for the Sun's meridian passage (Greenwich Meridian). It is necessary to take into account the fact that your longitude will probably be different from that of Greenwich when tracing out a sundial. For example, the Sun crossed the meridian in Paris at 11 h 55 min 25 s UT on 13 August 2000. What was the time of its meridian passage in Strasbourg (longitude −31 min 00 s)? Since Strasbourg is to the east of the meridian of Paris, the Sun crossed its meridian before that of Paris. We know that the longitude of Paris is −9 min 21 s. The difference in longitude between Paris and Strasbourg is therefore 21 min 39 s. The Sun crossed the meridian of Strasbourg at 11 h 55 min 25 s − 21 min 39 s = 11 h 33 min 46 s UT.

For Nantes (longitude +6 min 12 s), the Sun crossed the meridian on August 13 2000 at 11 h 55 min 25 s + 9 min 21 s + 6 min 12 s = 12 h 10 min 58 s UT.

Beware! The result is expressed here in Universal Time (UT). Since, in August, Summer Time (UT + 2 h) is in force, we must add 2 hours to our result in order to find the local (clock) time of the meridian passage.

There is a remarkable website (*www.bdl.fr*) run by the *Institut de Mécanique Céleste et de Calcul des Ephémérides* (IMCCE). It opens in an English-language version. Select "Sunrises, sunsets and transits of solar system bodies" and enter the date and your location in the appropriate windows. Select the celestial body (for our purposes, the Sun), and select "calculate". Geographical coordinates can be entered in a new window. Results are given in decimal minutes (perhaps not the most practical notation!).

Appendix H

Mottoes on sundials

This list is a selection of mottoes from more than 3500 which have to date been recorded on sundials. Any sundial worthy of the name should carry one!

From the Old Testament

A SOLIS ORTU USQUE AD OCCASUM
LAUDIBILE NOMEN DOMINI

The Lord's name be praised from the rising of the Sun unto the going down of the same (Latin, Psalm 112, verse 3). From a sundial on a private house in Le Mans (Sarthe).

From the New Testament

VIGILATE NAM NESCITIS QUA HORA

Be vigilant, for you never know at what hour ... (Latin). On a sundial dating from 1781 in the cathedral of Nevers (Nièvre). From the gospel according to St Matthew: "Vigilate ergo, quia nescitis qua hora Dominus vester venturus sit" (Be vigilant, for you never know at what hour your Master will come).

Christian motto

SANS LE SOLEIL JE NE SUIS RIEN
ET TOI SANS DIEU TU NE PEUX RIEN

Without the Sun I am nothing and you without God can do nothing (French). On a sundial by Zarbula, dated 1840, at Saint-Véran (Hautes-Alpes), on a barn in the hamlet of Le Raux.

Non-Christian motto

ΑΕΙ Ο ΘΕΟΣ ΓΕΟΜΕΤΡΕΙ

God is always doing geometry (Greek).

Philosophical motto

TEMS FAI CHANJA MADURA
OUBLIDA E MOURI

Time brings change, maturity, forgetfulness and death (Provençal). On a farm at the hamlet of Le Chambon, in the *commune* of Rémuzat (Drôme).

Moral motto

LES SOURIRES DONNÉS
NOUS REVIENNENT TOUJOURS

*The smiles we give always come back to u*s (French). On the church at Pontis (Alpes-de-Haute-Provence), on a sundial dated 2000.

Optimistic motto

POST TENEBRAS LUX

After darkness, light (Latin). On a private house in Bassenberg (Bas-Rhin) on a sundial dated MCMXCVIII (1998). From Job Chapter 17, verse 12: *Noctem verterunt in diem, et rursum post tenebras spero lucem—They (Job's 'comforters') change the night (misery) into day (hope), and after darkness I hope for light again.*

Epicurean motto

GENIESS JEDI SCHEENI STUND SE KOMMT
NUMME EINMOOL

Profit from each goodly hour, for it will never come again (Alsatian). On a sundial dated 1995 at a private house in Westhoffen (Bas-Rhin).

Mottoes on the passage of time

MIRA
L'HORA
PAS EL
TEMPS

Behold the hour—time is passing (Catalan). On a sundial dated 1988, on a private house in the village of Angoustrine (Angoustrine-Villeneuve-des-Escaldes, Pyrénées-Orientales).

DENEK DUTE KOLPATZEN AZKENAK DU HILTZEN

They all wound us, but it is the last that kills (Basque). On the south side of the church at Ossès (Pyrénées-Atlantiques). The motto is a translation into Basque of the well known Latin formula *Vulnerant omnes ultima necat.*

Astronomical motto

LE SOLEIL FAIT DV JOVR DES HEVRES
QVAND IL LVIT
MAIS DANS SON OCCIDENT
IL NOVS DONNE LA NVIT

The Sun makes the hours of the day when it shines, but in its setting gives us the night (French). On a chapel at the monastery of Notre-Dame-d'Orient at Laval-Roquecezière (Aveyron).

Motto on the working of sundials

SOL ME VOS UMBRA REGIT

The Sun guides me, the shadow you (Latin). At a farm at Sémeries (Nord), on a sundial dated "ANNO 1825".

Motto on the benefits of sunlight

HEP AN HEOL NON MANN EBET

Without the Sun there is no life (Breton). At Maël-Pestivien (Côtes-d'Armor), on the chimney of a private house in the locality of Ker Roland.

Humorous motto

TIBI QUOQUE MEDICE
ULTIMA VENIET

For you also, Doctor, the last (hour) will come (Latin). At Sains-Richaumont (Aisne), on a doctor's house.

Punning motto

SI SOL SILET SILEO

If the Sun falls silent, so do I (Latin). On a sundial dated 1992, on a private house at Saint-Vincent-les-Forts (Alpes-de-Haute-Provence).

Chronogrammatic motto

HORA FVGIT VELVT VMBRA
DIES LABVNTVR ET ANNI

Time flees away like a shadow; the days and the hours disappear (Latin). On the multiple sundial of the Abbey of Mont-Roland, near Dôle (Jura). This motto contains within its text the date of the construction of the sundial, if the Roman numerals are added together: hora fVgIt VeLVt VMbra DIes LabVntVr et annI: i.e. MDLLVVVVVVIII, or 1633.

Commercial and professional messages

NULLA DIES SINE PROTECTIONE MEO

No day without my protection (Latin). On an analemmatic sundial on the front of an insurance office at Aumetz (Moselle).

LIRE, POUR RESTER LIBRE

Read, to remain free (French). On a *méridienne* showing true time and mean time (bifilar) on a bookshop at Valréas.

Patriotic motto

VIV LA NATION

Long live the nation (French). On a sundial dated 1791, on a private house at Logis d'Antan, Crazannes (Charente-Maritime).

Political motto

J'INDIQUE LA FIN DE TES ENNUIS
MECANIQUES MAIS JAMAIS N'INDIQUERAI
LA FIN DES IMPOTS, TAXES, VIGNETTE ET
AUTRES GABELLES

I mark the end of your mechanical problems, but I will never mark the end of (freely translated) *income tax, VAT, road tax, and all the others* ... (French). At a petrol station in Aiguilles (Hautes-Alpes), on a sundial dated (with irony) "1965—AN VII DE L ERE REPUBLIGAULLIENNE" (Year VII of the 'RepubliGaullian' Era).

A personal/family motto

FERMETTA BON VIVERE FECIT

At La Fermette, life is good. On a private house called La Fermette at Monêtier-les-Bains (Hautes-Alpes), in rather unorthodox Latin.

Dedication

POUR TOI
LE TEMPS S'EST ARRÊTÉ
EN CE JARDIN DU SOUVENIR
POUR MOI
LE TEMPS S'ÉCOULE TOUJOURS
POUR HONORER TA MÉMOIRE

For you, time has ceased; in this garden of remembrance, for me, time goes on, to honour your memory (French). In the Garden of Remembrance at the civil cemetery at Braine (Aisne).

Commemorative text

TIME WILL NOT DIM THE GLORY
OF THEIR DEEDS

At the American Cemetery at Thiaucourt-Regniéville (Meurthe-et-Moselle), in memory of American soldiers killed in France.

Bibliography

F. W. Cousins, *Sundials*, ed. John Barker, London, 1969.

J. Evans, *The History and Practice of Ancient Astronomy*, ed. Oxford University Press, 1998.

G. Dohrn-van Rossum, *History of the Hour: Clocks and Modern Temporal Orders*, ed. University Chicago Press, 1996.

A. Gatty, *The book of sundials*, fourth edition, ed. Georg Bell and Sons, London, 1900.

M. Lennox Boyd, *Sundials: History, Art, Pelople, Science*, ed. Frances Lincoln, London, 2006.

R. Newton Mayall and Margarett W. Mayall, *Sundials: Their Construction and Use*, ed. Sky Publishing Corporation, Cambridge, USA, 1994.

D. Savoie, *La gnomonique*, ed. Les Belles Lettres, Paris, 2007.

A. Waught, *Sundials: Their Theory and Construction*, ed. Dover Publications, New York, 1996.

There are many Internet sites devoted to the theory and construction of sundials. Unfortunately, not all of these are written by competent persons, and one can find both the good and the bad! For anyone interested in finding out more, it is probably preferable to make contact with one of the many sundial societies listed below, who will be able to offer good practical advice.

Sundial Societies

http://www.sundialsoc.org.uk/ : British Sundial Society

http://sundials.org/ : The North American Sundial Society

http://cadrans_solaires.scg.ulaval.ca/v08-08-04/accueil/accueil.html : Quebec Sundials

http://www.astrosurf.com/saf/ : French Sundial Society

http://www.gnomonicaitaliana.it/ , http://quadrantisolari.uai.it/ : Italian Sundial Society

http://www.de-zonnewijzerkring.nl/eng/ : Dutch Sundial Society

http://relojesdesol.info/ , http://www.rellotgedesol.org/ , http://www.gnomonica.cat/ : Spanish Sundial Society

http://members.aon.at/sundials/ : Austria Sundial Society

http://www.societeastronomiquedeliege.be/ , http://www.gnomonica.be/ : Belgian Sundial Society

http://www.dgchrono.de/ : German Sundial Society

Sofware
http://pagesperso-orange.fr/blateyron/sundials/shadowspro/gb/index.html

Ephemerides
http://www.imcce.fr/imcce.php?lang=en : Institut de Mécanique Céleste et de Calcul des Ephémérides
http://aa.usno.navy.mil/data/ : U.S. Naval Observatory
http://ssd.jpl.nasa.gov/?horizons : NASA Jet Propulsion Laboratory

Glossary

Azimuth of the Sun   The horizontal coordinate of the Sun (cf. altitude). This is the angle between the direction of the Sun and due south. It is reckoned from due south along the horizon with positive sign from $0°$ to $+180°$ westwards, and with negative sign from $0°$ to $-180°$ eastwards.

Babylonian hours   Hours elapsed since sunrise.

Diurnal Arcs   These represent the curves described on a sundial for certain dates by the tip of the shadow of a style, normally at the summer solstice and the equinoxes. Diurnal arcs may be hyperbolae, circles, ellipses, parabolas or straight lines.

Ecliptic   The plane of the Earth's orbit around the Sun. To a terrestrial observer, it is the apparent path of the Sun through the sky during the year.

Equation of time   The difference between *true* solar time and *mean* solar time. The equation of time is the result of the axial tilt of the Earth and the eccentricity of the Earth's orbit.

Equinoxes   Those times in the year when the Sun crosses the Celestial Equator (normally 20 March and 23 September). The Sun is then at the zenith for places on the Equator. Night and day are both 12 hours long.

Gnomon   A stick (or similar) planted vertically in the ground. By extension, the study of the behavior of the shadow of a gnomon is known as gnomonics, the art of sundial design.

Hour angle   This is the angle, measured along the Celestial Equator from due south, between the meridian which passes through the Sun and the local meridian. It is reckoned with positive sign from $0°$ to $+180°$ westwards, and with negative sign from $0°$ to $-180°$ eastwards. A sundial in fact measures the hour angle of the Sun.

Italic hours   Hours elapsed since the previous sunset.

Latitude One of the geographical coordinates of a location (cf. longitude). Latitude is reckoned from the Equator, from $0°$ to $+90°$ in the northern hemisphere, and from $0°$ to $-90°$ in the southern hemisphere.

Longitude One of the geographical coordinates of a location (cf. latitude). Longitude is reckoned from the Greenwich Meridian, from $0°$ to $+180°$ westwards, and from $0°$ to $-180°$ eastwards. Longitude is often expressed in hours, minutes and seconds, $15°$ being equal to one hour.

Mean solar time True solar time corrected using the equation of time. Mean solar time is strictly local, and uniform.

Meridian, local A great circle on the celestial sphere passing through the northern point on the horizon, the North or South Celestial Pole, the zenith, and the southern point on the horizon. On the ground, it represents the north–south direction.

Motion, annual The motion of the Earth as it revolves about the Sun in one year.

Motion, diurnal The motion of celestial bodies as they curve westwards above the horizon. Diurnal motion results from the rotation of the Earth.

Obliquity of the ecliptic The angle $(23°\,26')$ between the Earth's axis of rotation and a line drawn through the center of the Earth perpendicular to the plane of the ecliptic. This also represents the maximum declination of the Sun at the summer solstice.

Solstices Those times in the year when the Sun is at its greatest distance from the Celestial Equator (normally 21 June and 21 December). The Sun is at the zenith at local noon for places on the Tropic of Cancer on 21 June, and for places on the Tropic of Capricorn on 21 December.

Style That part of a sundial which casts a shadow on the surface where hour lines and curves are marked. If it points towards the Pole Star, it is known as a 'polar' style. If it is perpendicular to the surface of the dial, it is known as an upright style. In the first case, the whole of the shadow of the style indicates the time. In the second, only the tip is relevant.

Sun, mean An imaginary Sun moving along the Celestial Equator and returning to the local meridian after exactly 24 hours.

Sun, true The Sun as actually observed.

Sundials
Analemmatic sundial: A normally horizontal sundial with a vertical style which moves according to the date. It shows solar time as a function of the Sun's azimuth: the direction of the shadow indicates the hour.
Altitude sundial: A portable sundial showing solar time as a function of the altitude of the Sun above the horizon.

Equatorial sundial: An inclined sundial with hour lines 15° apart. The dial table is parallel to the Earth's Equator.

Horizontal sundial: A horizontal sundial at a given location, with hour lines spaced according to a mathematical law and with a style pointing towards one of the celestial poles.

Polar sundial: An inclined sundial with parallel hour lines. The dial table is parallel to the axis of rotation of the Earth.

Vertical direct east sundial: A vertical sundial facing due east at a given location with (parallel) hour lines spaced according to a mathematical law and with a style, which points towards one of the celestial poles, parallel to the surface of the dial.

Vertical direct north sundial: A vertical sundial facing due north at a given location with hour lines spaced according to a mathematical law and with a style pointing towards one of the celestial poles.

*Vertical direct south (meridional) sundial:*A vertical sundial facing due south at a given location with hour lines spaced according to a mathematical law and with a style pointing towards one of the celestial poles.

Vertical direct west sundial: A vertical sundial facing due west at a given location with (parallel) hour lines spaced according to a mathematical law and with a style, which points towards one of the celestial poles, parallel to the surface of the dial.

Temporal hours   Also known as unequal or Biblical hours, these 'hours' were used by the ancients who divided the day into 12 parts irrespective of the season. The duration of the temporal hours varied with the date, and they were 60 minutes long only at the equinoxes (equinoctial hours).

True solar time   Time as shown by classic sundials. To be precise, true solar time is the hour angle of the Sun. It is strictly local and non-uniform.

Universal Time (UT)   Greenwich Mean Time + 12 hours.

Vernal Point (γ, First Point of Aries)   For a terrestrial observer, this represents one of the two points where the ecliptic crosses the Celestial Equator. The Sun reaches this point, crossing the Celestial Equator from south to north, on 20 March (vernal equinox). It reaches the equivalent point, when crossing the celestial equator from north to south, on 23 September (autumnal equinox).

Zenith   For a given location, the highest point in the sky, at 90° from the horizon. The direction of the zenith is defined by a plumb-line, which indicates the vertical of the location.

Index

Printing: Mercedes-Druck, Berlin
Binding: Stein+Lehmann, Berlin

Made in the USA
Middletown, DE
20 May 2018